INTRODUCTION TO
QUALITY ENGINEERING

Designing Quality into
Products and Processes

●

Genichi Taguchi

INTRODUCTION TO QUALITY ENGINEERING

Designing Quality into Products and Processes

●

Genichi Taguchi

Asian Productivity Organization

Available in North America,
the United Kingdom and Western Europe
exclusively from:

⌷⌷⌷QUALITY RESOURCES

White Plains, New York

and

AMERICAN SUPPLIER
INSTITUTE, INC.

Dearborn, Michigan

Some other titles published by the Asian Productivity Organization and available in North America, the United Kingdom and Western Europe exclusively from
QUALITY RESOURCES

Challenge of Asian Developing Countries: Issues and Analyses
Company-Wide Total Quality Control
Economic Engineering for Executives: A Common-Sense Approach to Business Decisions
Guide to Quality Control
How to Measure Maintenance Performance
Human Resource Development in Japanese Companies
Information Technology-led Development — Report on APO Basic Research
Japan's Quality Control Circles
The Japanese Firm in Transition
Japanese Management: A Forward-Looking Analysis
Japanese Management Overseas: Experiences in the United States and Thailand
Japanese Quality Control Circles: Features, Effects and Problems
Japanese-Style Management: Its Foundations and Prospects
Management by Objectives: A Japanese Experience
Modern Production Management: A Japanese Experience
100 Management Charts
Profitability Analysis: Japanese Approach
Quality Control Circles at Work
Reliability Guidebook
TPM: Total Productive Maintenance
White-Collar Knowledge Worker: Measuring and Improving Productivity and Effectiveness

Designed and Printed in Hong Kong by
NORDICA INTERNATIONAL LIMITED
for
Asian Productivity Organization
4-14, Akasaka 8-chome
Minato-ku, Tokyo 107, Japan

© Asian Productivity Organization, 1986

Eighth printing: 1990

ISBN: 92-833-1083-7 (Casebound)
ISBN: 92-833-1084-5 (Limpbound)

This book is available in North America, the United Kingdom
and Western Europe exclusively from:

QUALITY RESOURCES
One Water Street
White Plains, NY 10601
(914) 761-9600

and

American Supplier
Institute, Inc.
15041 Commerce Drive South
Dearborn, MI 48120
(313) 336-8877

"Sekkei-sha no tameno Hinshitsu Kanri: Hinshitsu Kogaku Gairon" (in Japanese) Copyright © Genichi Taguchi, 1983, Tokyo. Translated into English by the Asian Productivity Organization.

PREFACE

This textbook presents an outline of quality engineering (off-line quality control) as it affects product planning, research and development, design research, and production quality engineering. It elucidates the economic significance of quality problems and discusses methods of solving them. Written for staff personnel in planning, R&D, design, and production engineering departments, as well as managers in all departments, *Introduction to Quality Engineering* differs from other texts in that its methods for dealing with quality problems are based on economic calculations and product and process design. It assumes a knowledge of the fundamental techniques of experimental design.

On-line quality control (quality control during the production process) is also discussed briefly in Chapter 5 for comparison.

This text can also be used by a company to give two courses in quality control as outlined below. The schedules assume six hours of lectures and two hours of assignments per day.

Course 1: Six days

This course is a general treatment of quality control for designers and engineers. As many problems should be assigned as possible. If an introduction to experimental design and on-line quality control are added, the course can provide a good grounding in practical quality control techniques.

1st day: Chapters 1 and 2 and problems
2nd day: Chapters 3 and 4 and problems
3rd day: Chapters 5 and 6 (lectures)
4th day: Chapter 7 (lecture and problems)
5th day: Chapter 8 (lecture and problems)
6th day: Chapters 9 and 10 (lectures)

Course 2: Two days

This course presents an overview of quality control for managers and can be taught in 12 hours, 6 hours per day.

1st day: Chapters 1, 2, and 3
2nd day: Chapters 5, 6, and 9

If this text is used in an experimental design course or basic course, appropriate material can be selected from it to supplement the other lectures. In fact, this approach was used by the author recently in lecturing at major corporations in the United States.

The Author would like to take this opportunity to thank Dr. Don P. Clausing, Professor at the Massachusetts Institute of Technology(MIT), for his valuable advice on technical expressions, and Dr. Nuriel M. Samuel for his corrections by reviewing the English manuscript of this book which has been translated from the Japanese work.

Genichi Taguchi

Tokyo, April 1986

CONTENTS

SYMBOLS

A_0 : Functional failure loss, social loss

A : Loss by a defective found in production process

B : Measuring cost

C : Adjustment cost

f : Degree of freedom, number of squares linearly independant, rank of matrix attached to quadratic form corresponding to variation

F : Variance ratio

h : Interval

k : Coefficient of loss function, or number of levels

L_N : Orthogonal array of size N

n_e : Effective number of replication

q : Fraction of non-defective

S_T : Total variation, total sum of squares

S' : Net variation

S_A, S_B : Variation of main effects A, B

$S_{A \times B}$: Variation of interaction $A \times B$

Se : Error variation

V_T, V_e : Total variance, error variance

β : Coefficient or shrinkage factor

Δ_0 : Functional limit, customer's tolerance

$\Delta, \Delta_1, \Delta_2$: Tolerance limit

η ; Signal to noise ratio

ρ : Contribution ratio, percent contribution

o^2 : Mean square error, error variance

ω : ω-transformation or axis for minute analysis

1

VARIETY AND QUALITY

Functional Variability and Quality Problems

One facet of the current interest in alternative energy is the interest in generating electricity from wind power. Quite a number of suburban San Francisco homes use wind-powered generators, but they face a problem: on 15% of the days in the year, there is not enough wind to drive the generators; they have to be stopped, and no power can be obtained. The consumers' outrage if a commercial power company decided to shut down without warning on 15% of the days in the year can easily be imagined. The trouble with natural sources of energy is their variability. The same problem occurs to a greater or lesser degree in products and services provided by man. Machines break down, electric power fails, trains are late, roofs leak. The reason specifications are set is to prevent many of these problems. We shall begin by considering the loss caused by variability.

Although we shall focus on product variability, the same considerations apply to variability in services. The author defines the quality of an object as follows.

Quality is the loss a product causes to society after being shipped, other than any losses caused by its intrinsic functions.

This definition can be misunderstood on two points. One is that it appears to mean exactly the opposite of what is usually meant by quality. Some people hold that quality should be viewed as value. Value, however, is a subjective concept; everyone has his own idea of what constitutes value.

In elementary economics, value is defined as marginal utility. Economists were long puzzled as to why water, which is essential to life, should be cheap while diamonds, which are not, should be expensive. The demand for an article at a given price is the number of people who think it has value equal to or greater than that price. If the price is halved, the number of buyers will increase. In other words, demand is a function of price. If the supply increases, the price must be lowered to encourage demand. Oil is currently priced at $30 a barrel. If the price were increased to $100 a barrel, demand would fall sharply. The price is determined by the marginal utility -- if the consumers are ranked in order from those that value the article most to those that value it least, the marginal utility is the value assessed by the last person or group of people corresponding to the supply quantity. A certain camera company keeps its prices high by producing fewer of its product than could be sold at a profit, although it is thereby gradually losing market share.

Determining this subjective quality of value is a marketing and product- planning problem of vital importance for a company, but it is not an engineering problem. It is a human classification problem, a problem of segment size in market segmentation. This is why the author is opposed to treating quality questions as value questions.

Another problem is the meaning of loss. In the context of the definition of quality, loss should be restricted to two categories:

(1)   Loss caused by variability of function
(2)   Loss caused by harmful side effects

Examples of harmful side effects are familiar from medicine. Thalidomide was an excellent sedative, but its side effects caused terrible losses. In a television commercial a company claims that its motors always run at a constant speed. If this were true -- if they ran at exactly the same speed under all environmental conditions, year after year, despite wear and tear to parts and materials -- these motors would be perfect with regard to functional quality. If they also generated a great deal of vibration and noise, however, they would rank low in quality on the grounds of harmful side effects.

The Japanese <u>shinkansen</u>, the high-speed railway train, is highly acclaimed, but its functional quality suffers from variability -- it slows down in snow, and stops whenever there is an earthquake or damage to its overhead wiring -- and it causes side-effect problems in the form of excessive vibration and noise. Its functions represent a quantum leap over traditional rail travel, but it is plagued by quality problems. An article with good quality performs its intended functions without variability, and causes little loss through harmful side effects, including the cost of using it. If cost control is concerned with reducing the various losses that may occur before the product is shipped, quality control is concerned with reducing the two types of losses that it may cause to society after it is shipped.

Quality control is not, however, concerned with reducing the loss the product may inflict on society through its intrinsic functions. Besides such functions as aroma and taste, liquor has the function of intoxication. Untold numbers of people sustain losses from fights or accidents while under the influence of alcohol. Intoxication, however, is a function of liquor -- it is what liquor is for -- and it would be nonsense to manufacture nonintoxicating liquor because of the losses sustained through intoxication. The resulting product would not have the function of liquor.

The question of what functions society should allow products to have is a cultural and legal problem, not an engineering problem. Civilized countries allow even harmful products to exist and let people satisfy their curiosity by using them. A country that bans drugs and pornography cannot be said to be at the pinnacle of civilization. If we begin to discuss issues of utility and value, we are out of the realm of engineering and into the realm of cultural values. Engineers are free to debate these issues as individuals, but they must recognize that they are dealing with social questions that are outside the scope of quality control.

Children who become engrossed in television and neglect their studies may sustain major losses by failing to develop necessary skills. When a child grows up he may wish that the picture on the TV screen had been less appealing, so that he would not have become so addicted to it. Yet to advocate that TV programs should be made so unattractive that no one wants to watch them just so that children will not be tempted to neglect their studies would be to stand reason on its head. Quality control demands that the programming be as attractive as possible, and that the television sets malfunction as infrequently as possible. The question of whether to allow television is a cultural issue, and the question of whether to watch it or not is an individual issue. In summary, if quality control does not restrict itself to problems of loss caused by variability of function and unrelated harmful side effects, it will slip out of the domain of engineering into the psychological domain of cultural values. This is a barren arena of discussion, and every effort will be made to avoid it in this text.

Variety Problems and Quality Problems

If we focus on the manufacturing industry, we find that its activities fall into six stages.

(1) Product planning (which includes estimating the demand for a given function at a given price and setting the design life)
(2) Product design (designing the product to have the functions decided on in the planning stage)
(3) Production process design

(4) Production
(5) Marketing (including informing the market of the existence of the new product)
(6) Sales

Suppose the product is a man's dress shirt. Besides the different patterns and colors, this product comes in about a hundred different sizes. When buying a shirt, the customer will consider its color, pattern, material, size, and price. Pattern and color preferences depend on subjective values; one man may prefer blue, another pink. Some people may want a pattern that is distinctive. It is vital that the manufacturer know which colors and patterns will sell well, but the fact that blue sells better than pink does not mean that a blue shirt is better quality than a pink one. It is a question of taste. The market for blue shirts is different from the market for pink ones. There are also fashions in colors and patterns. Fashion is a phenomenon that causes changes in subjective value when there is no corresponding change in real value. Red may be in fashion one year, blue the next. Hemlines rise and fall. Since there is no real value difference, fashions change every year, often repeating themselves cyclically. The changes from vacuum tubes to transistors and then to integrated circuits were rapid, but these are not referred to as fashions because the transistor is better quality than the vacuum tube and the integrated circuit better quality than the transistor. Better quality means providing the same utility (function) with less loss to the consumer: with fewer failures, less power dissipation, and a longer service life. Color, pattern, and other subjectively-valued properties serve to classify products into varieties. Power dissipation, failure rate, service life, space occupied, and so on are quality items that bear on losses experienced by the consumer.

We have defined quality as the loss to society caused by a product after it is shipped (other than any losses caused by the product's intrinsic functions). A product with good quality will cause few such losses to society. No matter how much they like the color and pattern of a shirt, few people will buy it again if it soils or wrinkles easily, or accumulates static electricity, or causes skin rash. Goods are bought because of their utility and price. Utility includes external appearance and function; it determines what variety of product is being sold. At first, sales are controlled by variety and price. If a shirt is to be worn properly, however, it must be washed and ironed. At present, sending a shirt to the cleaner costs about ¥ 250 a time. Since the average shirt is apparently cleaned some 80 times during its useful life, the laundry expenses for one shirt come to about ¥ 20,000. If a new kind of shirt could be made that soiled and wrinkled only half as fast, the consumer would save ¥ 10,000 on laundry expenses. If it cost an extra ¥ 1,000 to make the new kind of shirt and the price was ¥ 2,000 higher, the manufacturer would gain ¥ 1,000 and the consumer ¥ 8,000. Nor would these be the only benefits; halving the number of cleanings would halve the amount of dirty water and noise produced by the cleaning process. In the currently popular phraseology, one would speak of a 50% reduction in pollution and a 50% saving in water, soap, and other resources.

Let us consider whether the size of a shirt is variety or quality issue. Shirts come in about 100 sizes for different neck size and sleeve length combinations. Neck sizes are at 1 cm intervals and sleeve lengths at 2 cm intervals. If the shirt is tailor-made, the customer's shoulder span, girth at the chest and waist, maximum upper arm girdle, back length, wrist girth, and other measurements are taken, paper patterns are made with the optimum allowance, and the material is cut and sewn. Tailor-made shirts, however, are not only expensive but also highly subject to variation because they are measured and sewn by hand, a process that depends on the skill of the tailor. Losses are generated because the shirt has to be tried on and cannot be obtained immediately. The trend is toward ready-to-wear or semi-tailored shirts and suits.

Since neck sizes are provided only in 1 cm gradients, a person with a 40.5-cm neck is forced to purchase a shirt that is 0.5 cm too loose or 0.5 cm too tight. If exactly the right size is available, the size loss is zero. The greater the deviation from the exact size, the greater the loss. The formula for calculating the loss will be given in the next chapter, but if the deviation is too great, the customer will probably not buy the shirt. Still, one cannot say that a shirt with a 41-cm neck is better in quality than one with a 40-cm neck. Providing sizes at intervals of 1 cm is a question of standardization. For the individual, not being able to find a shirt in the exact size is a quality problem. If the intervals between sizes were reduced, the customer's 40.5-cm neck size could be added to the selection, but by increasing the number of sizes (varieties) the manufacturer would be subdividing the market into smaller segments, causing further problems.

People in small market segments, such as people with large necks but short arms, often have trouble finding their sizes. Manufacturers do not make shirts in these sizes because the demand is too small, or retailers do not stock them because they are slow-moving. There need to be shops that specialize in unusual sizes. If the number of such special-size shops were 1/100 the number of ordinary retail shops, they could stay in business by selling their products at about 100 times the usual rate of turnover. There is a need for size specialization in the retail trade.

· Increasing the number of sizes is a question of increasing the number of varieties, but it is also a measure to counter the quality problem of the loss caused by the unavailability of exact sizes. The principal aim is to solve a quality problem through market segmentation. In summary, the variety problem is to provide a wide selection of colors, patterns, and sizes to meet people's individual needs. Deciding how much demand there will be for these colors, patterns, and sizes at a given price is a product planning problem. Deciding how to segment the market, and designing and pricing the varieties, are problems that involve all departments: planning, sales, and design. Once the varieties are decided, however, they will not involve the manufacturer in any complaints or product liability actions, because the customer knows the color, pattern, size, and price when he buys the article.

A few years ago the author visited South America in connection with a quality problem. Colombia had begun to industrialize about ten years earlier, but it was having problems with the quality of its products and wanted advice. In all countries, industrialization seems to start with textile manufacturing. There were a number of excellent textile factories in full operation.

Japan imports coffee and precious stones from Colombia, but Japanese imports are worth only about one-tenth as much as the machinery, steel, electrical appliances, and construction materials it exports in return. The Japanese trading firms were doing their best to find something to import from Colombia to reduce the trade imbalance, and their first candidate was textile products with Latin American colors and patterns.

When these products were actually imported, they sold well. Apparently the Latin American designs appealed to a large number of Japanese consumers. However, the consumers who bought these Colombian textile products found that the colors ran when they were washed. Department stores were flooded with complaints. These they passed on to the importer, but it was already too late. Most consumers had already made up their minds not to buy any more Colombian fabrics. When the author told Colombian government representatives and an industry representative these complaints, they said that Japanese consumers were too demanding. In substantiation, they cited the fact that North Americans were very good consumers for Colombian textile products. Color and pattern are variety issues, but fading in the wash is clearly a quality problem -- a problem of functional deviation. The author gave the Colombians the following bit of advice.

You won't succeed by imitating Japanese colors, patterns, and designs. You should continue to design and export products with Latin American features. Those colors, patterns, and designs appeal to lots of Japanese consumers. The Japanese market, however, is much more uniform than the American market. I have found that the price of a typical suit is only around $20 in department stores in the Mexican area of Los Angeles, *and at that price the consumer will be satisfied with low quality. In 1962, when I was at Bell Labs, a man who outranked me and earned around $3,000 a month said that he had bought the clothes he was wearing (shirt and trousers) for about $16, but he liked them and wanted to buy another pair. He thought that they had been made in Japan, and wanted to know the name of the company. The United States is an advanced country with a market stratified into many levels, from the high-quality level to the plain, inexpensive level. Accordingly, low-quality products will still sell if they are priced cheaply enough. Japan has not only a homogeneous population; it also has a homogeneous market. Check the quality of the goods you are competing with and try to close the quality gap. If you do that, I think you will find Japan an excellent market for your textile products.*

Besides showing the significance of variety and quality, this example indicates what types of specifications -- specification values in particular -- should be applied to variety and quality. People who buy textile products look at the color, pattern, design, size, and material. If the color runs or fades in the wash, that means the color the purchaser liked has changed; it is a change in function. Shrinking in the wash would be a similar problem. The Colombian producers had not adequately studied the quality problem of variability of function. The function itself, like the price, is clear when the article is purchased. Quality tends to be uncertain. The difficulty in the quality problem is that many qualities show up only after the article is purchased, when it is compared with an article made by a different manufacturer.

At first, how well a product sells depends on its function (e.g. its variety) and its price. If it is inferior in quality, however, its purchasers will lodge complaints or stop buying it, and all the effort and expense of developing the product, gearing up to produce it, finding a market for it, and so on will be wasted. Most quality problems become clear only afterwards. To avoid betraying the people's trust, a company must develop and produce products that are fully competitive in quality. Quality design in the product design stage is particularly important because, although variability can be reduced in the production stage, deterioration of the product or unsuitability for its environment cannot. All of these quality problems, however, can be dealt with in the design process.

PROBLEMS

1. List and describe the functions of a product with which you are involved. Select what you think to be the most important function

and consider what losses are caused by variability of that function.

2. What is the approximate cost of replacing a flourescent light that has gone dim or burnt out? Suppose the service life of a flourescent light is at present one year. If an improved type with a life of three years can be made, what is the monetary value of this quality improvement?

DISCUSSION

Variety, Quality, and Saleability

S (Student): When a product does not sell well, that is generally taken as an indication of inferior quality for the price, but could you explain in more detail what makes a product sell well?

G (Genichi): Many factors determine how well a product will sell: its utility, price, quality, and the efforts made to create a market for it. Advertising is important for new products, because no one will buy something he does not know about. There are also legal aspects. There are many articles -- drugs, for example -- that are forbidden by law despite a consumer demand for them. A legitimate company would not violate the law by manufacturing these products no matter how strong the demand.

S: So quality is one of the factors in sales.

G: Yes, but while improving quality improves a company's competitive position, it does not necessarily improve its total sales volume. Suppose, for example, that a company makes flourescent lights that last three times as long and that these cost no more than the old product. Obviously, people will buy only one-third as many. If the company had a 60% market share previously and its improved product captures 100% of the market, its sales volume will still fall to only 56% of the previous level.

Current sales revenues $\times$ 1/0.6 $\times$ 1/3
= Current sales revenues $\times$ 0.56

S: I've heard of an automobile parts manufacturer that caused itself a major loss by improving the quality of its product to the point where the product never failed, so retail sales of replacements fell to zero.

G: In December 1980 I met the head of the Michelin laboratories and asked him a similar question. Michelin tires are prized in Japan for their high quality, but I wondered if the longer life of the tires would not hurt future sales.

"At the present," he answered, "we are increasing our market share and our sales are growing. All the Japanese automobile manufacturers now buying our tires, and the Japanese monorails have used Michelin tires from the beginning. If improved quality causes our sales to drop in the future, we intend to develop other products that consumers want. We put quality first."

S: Quite a few companies take an indifferent attitude toward improving quality, particularly in their design work. Losing replacement sales because parts never fail and seeing sales drop because of longer service life are serious problems for a company. A firm that made wire fillets for spinning machines nearly went bankrupt after improving its products so that they lasted six times as long as before.

G: The purpose of the anti-trust laws is to foster competition. A few years ago there was a movement to break up a number of large American corporations: AT & T, IBM, Xerox, and some others. I had a chance to discuss this with an American economist at the time, and he said the government was out of its mind. All of those companies were in the leasing business. AT&T designed its switching systems and other equipment for a service life of forty years.

Why forty years? Because after about forty years, even if the hardware itself was still functioning, technological progress would have made it functionally obsolete. In other words, there is no need to make the service life unnecessarily long. Not that consumers would not be happy if products like light bulbs lasted a lot longer than they do. But when AT&T, IBM, and Xerox's products wear out, the firm has to replace them free of charge. It is to the firm's advantage to have its products last a long time. Companies in the leasing business only hurt themselves by making products that have short service lives or break down, so they try to make their products last.

S: I had not thought of the anti-trust laws as an important social system for improving product quality, but perhaps they are. But if the lease and rental business has a built-in incentive to improve quality, the anti-trust laws ought first to be applied to companies that sell outright.

G: Exactly. Of course, a company is free to turn out products of the worst quality imaginable as long as they do not break the law.When the consumers find out about quality problems, they are supposed to react by complaining or boycotting the company. Although one of a company's aims should be to make products that last longer without costing more, even if that reduces its volume of sales, monopolistic firms tend to stop making those kinds of improvements. That is why we have the anti-trust laws, but firms that lease their products will go on improving them even after they have a monopoly.

S: I am beginning to appreciate the difficulties involved in forecasting sales.

Can Variety and Quality Be Distinguished?

S: Can you explain the difference between quality and variety in a little more detail?

G: We buy products when we think the price is equal to or less than the utility -- what they can do. A product without utility or with inadequate utility is defective. The expected utility of a product is a question of variety. If the product ceases to function or functions inadequately, that is a quality problem.

S: That explains why service life is a quality problem. But for a car, say, is having a maximum speed of 100 km/h or 150 km/h a question of variety or of quality?

G: Think in terms of market segmentation. Let one market consist of people with no use for speeds in excess of 100 km/h and another market consist of people who think they want to go faster than 100 km/h. People in the first market may or may not buy a car that can do 150 km/h. People in the second market, however, will definitely not buy a car that can only do 100 km/h, because it lacks a function they require. In other words, the functions demanded in the two markets are different.

S: As far as speed is concerned, the car that does 150 km/h satisfies the functional requirements in both markets, so it can sell in both of them. I guess that makes it a more versatile car, but does not make it a better quality car.

G: That's right, but as I said before, there are cases in which the distinction between variety and quality is not clear. I would not be against saying that maximum speed is a quality characteristic which different people value in different degrees. What I am against is calling everything a quality characteristic. Although some things could be called either, the distinction ought to be made clear in some cases .

S: I don't think many quality control people would disagree with you there. But it is clear that even when we are talking about characteristics that are not of direct utility, like instant warm-up in a television set, we are talking about quality.

G: If the average television set is turned on four times a day and has a design life of eight years, the time the buyer loses by having to wait about 10 seconds for the picture to appear instead of getting it instantly can be calculated as:

$$10 \text{ sec} \times 4 \text{ times} \times 365 \times 8 = \text{approx. } 32 \text{ hours}$$

If the set is watched by an average of two people at once, viewers lose a total of 64 hours during its service life. If we calculate this loss at ¥ 10 per person per minute, it comes out as ¥ 38,000. A set that gives an instant picture has to draw a certain amount of current even when turned off, so to see which is better, you have to calculate the cost of the electricity over eight years. Having to wait for the picture to appear is not the utility expected of the television set; it is a loss endured to obtain the utility. It is purely a question of quality, not variety.

If quality is defined as the loss a product causes to society after it is shipped, then automobile exhaust, air conditioner noise, the toxic gas from incinerating plastic containers, and all other pollution problems that cause third-party losses are quality problems. Since the third party is not the purchaser of the product, he does not receive any of its utility; all he gets is the loss.

Improvements in quality are always beneficial to society. Variety, however, is something that cannot be changed even if its utility is accompanied by a loss. To return to the examples in the text, children who become addicted to television suffer a loss, as do grown-ups who become alcoholics. Getting intoxicated, however, is the utility of alcoholic beverages. If an alcoholic beverage were made that spared people the harm of intoxication, it would no longer be classified as an alocoholic beverage. Its utility would have changed; it would have become a different breed of product. The harmful, hangover-causing aldehyde content of the beverage, however, and whether it contains preservatives, or whether it will spoil if it does not contain preservatives, would be quality questions.

What about taste? Having a different kind of taste is a variety question, not a quality question. Having taste that is not necessarily different but that everybody finds disagreeable, however, is a quality question. If rice is badly prepared and some of the grains are only half cooked, everyone will agree that it tastes bad. This is a quality question, not a question of personal preferences; but at fine levels of discussion it can be difficult to distinguish between variety questions and quality questions.

Extra functions are basically variety questions. An example of an extra function would be equipping a television set to play video cassette tapes. Accessories are also extra functions. In many questions of variety, the value judgments are subjective, with wide disparities between different people's valuations. The only solution is to move toward a greater variety of products.

Clothing must be made in a variety of patterns and colors. There is a tendency for people to want different patterns and colors, which shows how complex the question of variety can become. A company must decide what allocation of its production among different patterns will increase its market share. The situation is even more extreme in the production of art objects: the same object must not be made twice.

It is desirable to satisfy different people's different tastes, but the mere fact of satisfying tastes does not in itself signify progress. As society becomes more affluent, values will necessarily become more diverse. The trend toward greater variety is a historic trend.

I think it is important to shape society in the directions of greater variety and higher quality, which includes less pollution with regard to third parties.

S: I can see why it can be hard to make the distinction in individual cases, but I think I have the general idea.

2 | VARIABILITY LOSS AND TOLERANCE

Ready-to-Wear Clothing and the Size Problem

One of my interests is distribution surveys of body sizes and foot sizes to set standard sizes for ready-to-wear clothing and shoes. Apparently only 30% of Japanese consumers are able to buy shoes that exactly fit their feet, and 50% say they have trouble finding clothes sizes that fit, although this varies depending on the type of clothing. If deciding what sizes to provide is the task of product planning and the purpose of size design, then a country that supplies only 50% of its consumers with the right size clothing and only 30% with the right size shoe can hardly be said to be performing satisfactorily in these fields. In this chapter we will consider the setting of standard values, in particular the spacing of sizes, by examining the problem of clothing size standards. We shall use the example of men's dress shirts, which is particularly simple.

Tolerance and Interval

Men's dress shirts are currently sold in a maximum of about 100 sizes, a size being defined by two measurements: the neck measurement and sleeve length (the length from the center line of the back to the edge of the cuff). Neck sizes are provided at 1 cm intervals, and sleeve lengths at 2 cm intervals. A shirt with a 40-cm collar will just fit a person with a 40-cm neck. A person with a 40.5-cm neck must buy either a 40-cm neck which is 0.5 cm too tight, or a 41-cm neck which is 0.5 cm too loose. In the present size system, there is a tolerance of:

$$\pm 0.5 \text{ cm} \qquad\qquad\qquad ......(2.1)$$

If a tight collar is less bearable than a loose one, then the tolerances can be set asymmetrically on the plus and minus sides, as follows:

$$-0.3 \text{ cm}$$
$$+0.7 \text{ cm} \qquad\qquad\qquad ......(2.2)$$

Are the current size intervals of 1 cm for necks and 2 cm for sleeves reasonable? A person who cannot find his exact size experiences a loss. Since only 30% of the consumers can find shoes that fit perfectly, 70% are experiencing a loss because their exact size is not available. With all the products on the market, it is strange that a consumer should be unable to find one that satisfies his requirements. Providing products to satisfy the needs of as many individual consumers as possible is both a problem of standards and a measure of the progress of civilization. It is not, however, possible to satisfy everyone's individual needs; for shirts, that would mean having the cloth, the size, and the design all made to individual order, and the price would be prohibitive for most people. Accordingly, sizes can be standardized, with attendant cost savings, provided the deviation from the optimum size is small enough that the attendant loss is also small. The question is the size tolerance -- the interval between sizes.

Determining the Neck Size Interval

Many measurements have to be made to determine the size of a shirt: neck, sleeve length, shoulder span, sleeve length measured from the shoulder, girth at the chest and waist and around the hips, height, maximum upper arm girdle, wrist size, and more. For each of these measurements a distribution can be obtained by measuring a sample of the consumer population. It would be convenient if the scatter in these distributions were small, but in fact it is large, so many different sizes have to be made to satisfy consumers' needs.

Let m be the exact size for a certain measurement. The question is at what deviation from size m the consumer will refuse to buy the shirt. This deviation is the consumer tolerance.

For neck size, if a person will not buy a shirt that is Δ_1 cm smaller than his exact size, no matter how much he likes the color and pattern, then Δ_1 is his lower tolerance. If he will not buy a shirt that is Δ_2 larger than his exact size because it is too loose, then Δ_2 is his upper tolerance. If different people have different tolerances with respect to the same measurement, the tolerance is set at the point at which 50% of the consumers will not buy, called the *LD50* point. (In drug testing, this designates the dose at which the live/die ratio is 50:50.) The consumer tolerance is expressed as an ordered pair ($-\Delta_1$, $+\Delta_2$).

With y the size of the shirt, m the buyer's exact neck size, and $L(y)$ the loss due to the difference between y and m, $L(y)$ can be expanded in a Taylor series around m as follows.

$$L(y) = L (m + y - m)$$

$$= L (m) + \frac{L' (m)}{1 !} (y - m) + \frac{L'' (m)}{2 !} (y - m)^2 + \quad ......(2.3)$$

By assumption, $L(m) = 0$, and since $L(y)$ is minimal at $y = m$, $L'(m)$ is zero. Accordingly, the third term of the above expansion of the loss function is the main term, and the loss can be approximated as:

$$L(y) \doteqdot k (y - m)^2 \qquad ......(2.4)$$

Formula 2.4 for the loss function contains the unknown coefficient k. To find k, we need information about the losses D_1 and D_2 caused by exceeding the tolerances Δ_1 and Δ_2. In the present case, if the tolerances are exceeded, the shirt must be tailor-made (or adjusted), and the losses include the additional tailoring, the delay, the cost of mailing the shirt or going to get it, and other social costs. Let us assume that D_1 and D_2 are both ¥ 4,000. If all the available shirts deviate by more than Δ_1 or Δ_2 from the exact size m, the shirt must be tailor-made, notwithstanding the ¥ 4,000 loss. Accordingly,

$$\text{When } y < m, \ k_1 = \frac{4000}{\Delta_1^2} \qquad ......(2.5)$$

$$\text{When } y > m, \ k_2 = \frac{4000}{\Delta_2^2} \qquad ......(2.6)$$

Suppose that $\Delta_1 = 0.5$ cm and $\Delta_2 = 1.0$ cm. Then

$$k_1 = \frac{4000}{0.5^2} = 16000 \qquad ......(2.7)$$

$$k_2 = \frac{4000}{1.0^2} = 4000 \qquad ......(2.8)$$

The loss function $L(y)$ is accordingly given by:

$$L(y) = \begin{cases} 16000\,(y-m')^2 & (y < m') \\ 4000\,(y-m')^2 & (y > m') \end{cases} \qquad \text{......(2.9)}$$

The value of m' can be determined as follows.

If neck sizes are provided at 1-cm intervals, a person with a 40.3-cm neck must choose a shirt with a neck size of either 40.0 cm or 41.0 cm. Which choice will entail the greater loss? This is partly a matter of preference, but if we compare according to formula 2.9,

$$L(y) = \begin{cases} 16{,}000 \times 0.3^2 = ¥\,1{,}440 \\ 4{,}000 \times 0.7^2 = ¥\,1{,}960 \end{cases} \qquad \text{......(2.10)}$$

This person will accordingly prefer the slightly-too-tight 40-cm size. The two losses in formula 2.9 would be equal if the person had a 40.33-cm neck.

When setting the size interval, it should be assumed that all consumers will choose the smaller loss. Next the average loss can be considered. If we assume that people's neck measurements are uniformly distributed in the interval between 40.0 cm and 41.0 cm (which is equivalent to approximating the distribution function by a bar graph), and call the average loss L, then:

$$L = 16000 \times \int_{40.00}^{40.33} (x-40.0)^2 \, dx + 4000 \times \int_{40.33}^{41.00} (x-41)^2 \, dx$$

$$= 16000 \times \frac{0.33^3}{3} + 4000 \times \frac{0.67^3}{3} \fallingdotseq ¥\,593 \qquad \text{......(2.11)}$$

The loss due to having to make and sell one extra size can be assumed to be nearly equal to the difference between the retail value and the proportional cost, in this case ¥ 1,800. Under these conditions, 1-cm neck size intervals are too small; the added manufacturing and selling cost outweighs the average loss to the consumer. Yet if the interval is increased to 2 cm, the loss is:

$$L = 16000 \times \int_{40}^{40.67} (x-40.0)^2 \, dx \times \frac{1}{2} + 4000 \times \int_{40.67}^{42.00} (x-42)^2 \, dx \times \frac{1}{2}$$

$$= ¥\,2{,}370 \qquad \text{......(2.12)}$$

Since this value is larger than ¥ 1,800, an interval of 2 cm is too wide. The ideal interval is the one at which the consumers' loss and the manufacturers' and sellers' loss are equal. This value (Δ) comes out to be 1.8 cm. Under the assumptions made above, the interval of 1 cm should be increased by 0.8 cm.

The same type of calculation can be carried out for all the other measurements. In general, the optimum size interval will not be the same as the consumer tolerance range. For ready-to-wear products, the value of ¥ 1,800 (the retail value minus the wholesale value) includes such selling costs as inventory and interest costs. Although there is no harm in using these costs to determine the interval, the cost of maintaining stocks in sizes for which there is only small demand is very high. If D is the loss caused by having to keep all sizes in stock, then D involves factors such as inventory cost that vary with the size of the demand. Borrowing a term from economics, we can call the limit demand at which it is profitable to maintain a stock the marginal demand.

For some commodities, most of the increased cost of producing more varieties falls on the producer in the form of the cost of extra patterns and extra tools and setup changes. In these cases, however, the same formula can still be used.

Functional Limits and Tolerances

Let us consider the problem of fitting a pane of glass into a window frame, whether it is in a building or in an automobile. Both the pane of glass and the frame have specified dimensions and tolerances. Here we will assume that the specifications for the frame are nominal and find the tolerances for the window glass. The dimensions of each frame are different, but we need considering only the nominal value as $LD50$ in the frame dimensions.

Suppose that the dimensions of the glass are apt to vary. If they are too large, the glass will not fit in the window; if they are are too small, it will fall out. Let m_1 be the fall-out dimension, and m_2 the too-big dimension. There are dimensions like these in both the width and height directions, but let us limit ourselves to the width direction. Let m be the mean of m_1 and m_2, and Δ_0 be half the difference between them.

$$m = \frac{1}{2}(m_1 + m_2) \qquad\qquad ......(2.13)$$

$$\Delta_0 = \frac{1}{2}(m_2 - m_1) \qquad\qquad ......(2.14)$$

The consumer tolerance range for the width of this pane of glass is therefore:

$$m \pm \Delta_0 \qquad\qquad ......(2.15)$$

The tolerance specification for the pane of glass, however, will not be $\pm \Delta_0$. Although $\pm \Delta_0$ is the tolerance for the user of the glass, it is not the tolerance observed by the manufacturer who cuts the glass or the retailer who sells it. This is because the loss that occurs when the pane of glass is purchased and then taken to the installation site and found not to fit the window is due to the need to have the glass cut again, or to purchase a different size of glass. This loss includes the cost of transporting the glass from the retail store to the window site and back again. The pane of glass may cost only ¥ 300, but the round-trip transportation might cost several times that.

Let y be the width of the pane of glass and $L(y)$ the loss function. $L(y)$ can be expanded in a Taylor series around the median value m, which is the most desirable value, as follows:

$$L(y) = L (m + y - m)$$

$$= L (m) + \frac{L' (m)}{1 !} (y-m) + \frac{L'' (m)}{2 !} (y-m)^2 + \cdots \cdots \quad \ldots\ldots(2.16)$$

When the dimension is the ideal value m, the loss $L(m)$ is zero, and the first derivative $L'(m)$ should also be zero, so the above series actually begins with its third term. If we approximate by omitting the higher-order terms, we have:

$$L(y) \fallingdotseq k (y - m)^2 \quad \ldots\ldots(2.17)$$

Let A_0 be the average loss incurred if the pane of glass turns out to be too wide or too narrow and has to be recut. This loss of A_0 occurs when $|y - m|$ is Δ_0, so the coefficient k can be estimated as follows:

$$k = \frac{A_0}{\Delta_0^2} \quad \ldots\ldots(2.18)$$

The tolerance Δ of the pane of glass can be determined if we know the loss A incurred at the place where the glass is cut, when it does not conform to the tolerance. The loss function is:

$$L = \frac{A_0}{\Delta_0^2} (y - m)^2 \quad \ldots\ldots(2.19)$$

Substituting A for the left side and solving for $\Delta = |y - m|$ gives:

$$A = \frac{A_0}{\Delta_0^2} \times \Delta^2 \quad \ldots\ldots(2.20)$$

This results in the formula:

$$\Delta = \sqrt{\frac{A}{A_0}} \times \Delta_0 \qquad\qquad ......(2.21)$$

Let the loss A_0 when the glass turns out to be the wrong size at the installation site be ¥ 1,500, the tolerance range Δ_0 be 3 mm, and A be ¥ 300. The tolerance is then given as follows.

$$\Delta = \sqrt{\frac{A}{A_0}} \times \Delta_0$$

$$= \sqrt{\frac{300}{1500}} \times 3$$

$$= 1.34 \text{ (mm)} \qquad\qquad ......(2.22)$$

Accordingly, the specification range for this pane of glass should be set at $m \pm 1.34$ (mm), and panes of glass outside this range should not be sold.

Loss Function

Assume that for the example in the preceding section, the actual dimensions of a set of panes of glass cut at the factory are as follows (the values below being the deviations from the target value).

0.3, 0.6, −0.5, −0.2, 0.0, 1.0, 1.2, 0.8, −0.6, 0.9, 0.0, 0.2, 0.8, 1.1, −0.5, −0.2, 0.0, 0.3, 0.8, 1.3

To find the loss due to variability, we find the mean squared deviation from the target value and substitute it into the loss function. For compactness of terminology, the mean squared deviation from the target value will be called the variance and denoted by σ^2.

$$\sigma^2 = \frac{1}{20} \left[0.3^2 + 0.6^2 + \cdots\cdots + 1.3^2 \right]$$

$$= 0.4795 \text{ (mm}^2) \qquad\qquad ......(2.23)$$

Using this mean value in formula 2.19 gives:

$$L = \frac{A_0}{\Delta_0^2} \times \text{mean squared deviation from target}$$

$$= \frac{1500}{3^2} \times 0.4795$$

$$= ¥79.9 \qquad \qquad ......(2.24)$$

Hereafter S denotes variation (sum of squares) and f degree of freedom.

The average dimension for these panes of glass is 0.365, slightly on the large side. The situation can be evaluated by constructing an analysis-of- variance table. Since:

$$S_T = 0.3^2 + 0.6^2 + \cdots + 1.3^2 = 9.59 \, (f=20) \qquad .....(2.25)$$

$$S_m = \frac{(0.3+0.6+\cdots\cdots+1.3)^2}{20} = \frac{7.3^2}{20} = 2.66 \; (f=1) \qquad ......(2.26)$$

$$S_e = S_T - S_m = 9.59 - 2.66 = 6.93 \; (f=19) \qquad ......(2.27)$$

The analysis-of-variance table is as follows. See reference (1).

Table 2.1 Analysis of Variance

Source	f	S	V	F_0	S'	$\rho\;(\%)$
m	1	2.66	2.66	7.29	2.295	23.9
e	19	6.93	0.365		7.295	76.1
T	20	9.59	0.4795		9.590	100.0

Table 2.1 indicates that the mean value is too large in relation to the median value of the tolerance interval, resulting in an increased variance and an increased loss due to variability. Frequently it is a simple matter to adjust the mean value to the target value. If this is done, the variance σ^2 should become roughly equal to the error variance V_e in the table. The value of the loss function would then be:

$$L = 166.7 \times 0.365 = ¥60.8 \qquad ......(2.28)$$

In comparison with the results of formula 2.24, this is a quality improvement per pane of glass of:

$$79.9 - 60.8 = ¥19.1 \qquad ......(2.29)$$

If a hundred thousand panes are produced per month, the monthly quality improvement will be ¥1.91 million.

The effort to adjust the mean value to the target value does not require any special tools. All it requires is to use a rule, or to compare the product value with the target value, from time to time or every time. If the pitch interval used on the rule in making the comparison is changed, then both the mean and the variance can be changed. This is a question of calibration cycle in quality control. Data should be obtained for a variety of calibration cycles and a table like Table 2.1 constructed, or the loss function computed, to decide the optimal calibration cycle.

Inadequate Process Capability

We have been studying the setting of tolerances in product specifications, and found that three parameters are necessary. In what follows, for convenience, we shall assume that the upper and lower tolerances are the same. The calculations are derived in the same way when they are different.

Δ_0 = consumer (customer) tolerance
A_0 = cost to society when the tolerance is exceeded
A = cost to the manufacturer when the product is rejected

We have yet to give a detailed explanation of the parameter A. If A is known, then the manufacturer's tolerance (the tolerance noted on the drawings) Δ is given by the following formula.

$$\Delta = \sqrt{\frac{A}{A_0}} \times \Delta_0 \qquad\qquad ......(2.30)$$

Certain difficulties arise with the value of A, however, when there is inadequate process capability. Regardless of the tolerance, if a rejected article can be reworked into a satisfactory article at an average cost of A, then A is a repair cost and formula 2.30 can be applied as it stands. In some cases, however, articles that do not satisfy the tolerance limit Δ have to be scrapped, so when a unit is rejected there is a chance that the new unit manufactured to replace it will also be rejected. Let q be the proportion of good articles for a given tolerance Δ, so that q is a function of Δ. If the pass rate is only 50% for a certain value of Δ, there is only a 50-50 chance that a replacement will be acceptable. The cost of making an unacceptable article then doubles from A to $2A$.

In general, in a process in which defective products have to be scrapped, the tolerance Δ indicated on the drawings is the solution to the following equation.

$$\Delta = \sqrt{\frac{A/q}{A_0}} \times \Delta_0 \qquad\qquad ......(2.31)$$

Since q is a function $q(\Delta)$ of Δ, the above equation is a nonlinear functional equation.

If the characteristic value y is normally distributed, with mean m at the target value and standard deviation σ, $q(\Delta)$ is given by the following equation.

$$q(\Delta) = \int_{m-\Delta}^{m+\Delta} \frac{1}{\sqrt{2\pi}\sigma} e^{-\frac{1}{2\sigma^2}(y-m)^2} dy$$

$$= \int_{-\Delta/\sigma}^{\Delta/\sigma} \frac{1}{\sqrt{2\pi}} e^{-\frac{t^2}{2}} dt \qquad \qquad ......(2.32)$$

The tolerance Δ is accordingly the solution to the following equation.

$$\Delta = \sqrt{\frac{A}{A_0 \times q(\Delta)}} \times \Delta_0 \qquad \qquad ......(2.33)$$

For instance, suppose the consumer tolerance limit is $m \pm 200$ (μm), the loss A_0 caused by exceeding the tolerance limit is ¥ 8,000, the factory cost of the article is ¥ 300, and the standard deviation of the manufacturing process is 40 μm. According to formula 2.30, the tolerance Δ on the drawing should be:

$$\Delta = \sqrt{\frac{A}{A_0}} \times \Delta_0 = \sqrt{\frac{300}{8000}} \times 200 = 39 \qquad (\mu m) \qquad ......(2.34)$$

If the drawing tolerance Δ is 39, then since the standard deviation σ of the characteristic value is 40 μm, some out-of-tolerance articles will be produced. If the consumer is an assembly plant and the supplier does not have a monopoly, the consumer is likely to be unyielding toward the tolerance. Accordingly, efforts need to be made to improve the process capability. Here, however, we are discussing tolerances for products sold to general consumers.

$$q(\Delta) = \int_{-\Delta/40}^{\Delta/40} \frac{1}{\sqrt{2\pi}} e^{-\frac{t^2}{2}} dt \qquad \qquad ......(2.35)$$

Since Δ is unknown, $q(\Delta)$ is also unknown. A good way to obtain the tolerance Δ is by successive approximations.

First, using formula 2.30, we find the first approximation to Δ as in equation 2.34. The first approximation to $q(\Delta)$ is therefore:

$$q(39) = \int_{-39/40}^{39/40} \frac{1}{\sqrt{2\pi}} e^{-\frac{t^2}{2}} dt \qquad \qquad ......(2.36)$$

The integral on the right side will be abbreviated $F(39/40)$ below. $F(t)$ denotes the integral of the standard normal distribution between the limits $\pm t$; values are given in tables. From one of these tables,

$$q(\varDelta) = F(0.975) = 0.670 \qquad \qquad ......(2.37)$$

We can substitute this into equation 2.33 to obtain a second approximation.

$$\varDelta = \sqrt{\frac{300}{8000 \times 0.670}} \times 200 = 47.3 \qquad \qquad ......(2.38)$$

Then a second approximation to $q(\varDelta)$ can be obtained.

$$q(\varDelta) = F\left(\frac{47.3}{40}\right) = 0.762 \qquad \qquad ......(2.39)$$

Substituting this into equation 2.33, we obtain a third approximation to $\varDelta$.

$$\varDelta = \sqrt{\frac{300}{8000 \times 0.762}} \times 200 = 44.4 \qquad \qquad ......(2.40)$$

The third approximation to $q(\varDelta)$ is therefore:

$$q(\varDelta) = F\left(\frac{44.4}{40}\right) = 0.733 \qquad \qquad ......(2.41)$$

The fourth approximation to $\varDelta$ is $\varDelta = \sqrt{\dfrac{300}{8000 \times 0.733}} \times 200 = 45.2$

$$......(2.42)$$

If we terminate the approximation computations at this point, the value of the tolerance is:

$$\varDelta = 45 \ (\mu m) \qquad \qquad ......(2.43)$$

Compared with equation 2.34, this value is only a little wider. Even with this new tolerance, the percent defective will be:

$$1 - 0.733 = 0.267 = 26.7\% \qquad \qquad ......(2.44)$$

This means that it is fundamentally incorrect to set wide tolerances on the basis of process capability considerations. If the expected defective rate is less than 20%, the tolerance can be set by formula 2.30, and if the defective rate exceeds 20%, the primary problem is to improve the process capability.

Accordingly, we do not need to consider process capability, but we do need to consider the loss A incurred when a tolerance is exceeded. If sophisticated process capability goes hand in hand with high production costs, then it is customary to consider that in the value of A.

Destructive Inspection

Let us consider how to set tolerances for characteristics such as bonding strength, service life, tensile strength, compressive strength, and melting point that can only be measured destructively. Many of these have one-sided specifications, but many others have specification limits on both sides for safety or other reasons. Leaving one-sided specifications for a later chapter, we will here consider the problem of setting tolerances for a hardness characteristic with target value m.

Five hundred units of the product are produced per hour, eight hours a day, 250 days a year. Post-shipment troubles will occur if the hardness of the product is either too high or too low. The tolerance Δ_0 is ± 15 on the Rockwell hardness scale, the post-shipment loss A_0 caused by exceeding this tolerance is ¥ 60,000, and the loss caused by a defective article at the factory is ¥ 500. Given adequate process capability, the tolerance to mark on the drawings can be found as follows:

$$\Delta = \sqrt{\frac{A}{A_0}} \times \Delta_0 = \sqrt{\frac{500}{60000}} \times 15 = 1.4 \qquad ......(2.45)$$

Accordingly, the tolerance on the factory drawings should be $m \pm 1.4$ on the Rockwell hardness scale.

The factory will then control hardness to try to keep the production process operating normally. Process malfunctions may cause the hardness to vary outside its specification limits, but if such events occur only rarely, formula 2.30 can be used to set the tolerances.

What if defectives are produced even when the process is normal? The answer given in the preceding section was that if your competitors' process capability is adequate and your own is inadequate, you will lose out in the market and your products will not sell. But what if all manufacturers' products contain a significant fraction of defectives?

Suppose you have measured the Rockwell hardness of ten of your own products and found the following deviations from the target value *m*:

$$1.8, \ -3.1, \ 0.0, \ 2.5, \ 1.0, \ 1.8, \ -1.6, \ 3.2, \ 2.4, \ -2.0$$

The variance o^2 is calculated as below. There are ten degrees of freedom.

$$o^2 = \frac{1}{10} \, [1.8^2 + (-3.1)^2 + \cdots\cdots + (-2.0)^2 \,] = 4.59 \qquad \text{......(2.46)}$$

The coefficient k of the loss function is:

$$k = \frac{A_0}{\Delta_0^2} = \frac{60000}{15^2} = 266.7$$

Hence the loss L due to variability of hardness is:

$$L = 266.7 \times 4.59 = \yen 1{,}224 \qquad \text{......(2.47)}$$

This value is much larger than the factory cost of ¥ 500, but that does not necessarily mean that this company's products will find no takers. It only means that in the quality represented by variability of the hardness, there is a loss that is much larger than the production cost. To the company, these figures say that the production process needs to be gotten under control within the 1.4 tolerance, if that can be done for a cost of ¥ 500 or a little more per item.

Of course the urgent need is to improve the process so that it produces only very few defectives under normal conditions. This is a task for off-line quality control at the factory, a problem to be dealt with by the engineering staff. The quality of even characteristics that require destructive testing can be controlled, and production processes can be improved. It is wrong to compromise on the tolerance limits because of lack of process capability. It would be far better to estimate the cost of enhancing the process capability and request a price adjustment.

As discussed in the next section, another way to improve quality is to eliminate defectives by 100% inspection. The inspection process can be thought of as part of the production process, with the name "screening." When the cost of improving process capability is higher than the cost of 100% inspection, companies have been known to leave the process capability as is and simply screen out all the defectives. Similarly, in cases of destructive inspection, a company might decide that although the process is turning out too many defectives even when it is operating normally, improving the process would cost more than the loss from shipping the defectives.

Cost-loss comparisons like this, however, may be inaccurate because even products within the specification limits cause a loss if they deviate from the target value. Suppose, for example, that improving the process would involve an annual cost of ¥180 million (interest, depreciation, running cost), but would reduce the standard deviation of the hardness to one-fourth of the tolerance range of 2.8. (The tolerance range is twice the tolerance value.) The cost per unit would be:

$$\frac{\text{Per-annum cost increase}}{\text{Annual production volume}} = \frac{180,000,000}{500 \times 8 \times 250} = ¥180 \qquad \text{......(2.48)}$$

The improvement in quality would be ¥1,093, since the loss would be reduced from ¥1,224 to ¥131:

$$L = 266.7 \times \left(\frac{2.8}{4}\right)^2 = ¥131 \qquad \text{......(2.49)}$$

Since a quality improvement of ¥1,093 is achieved at a unit cost of ¥180, the gain from the improvement is:

$$1,093 - 180 = ¥913 \qquad \text{......(2.50)}$$

In annual terms, this is a gain of:

$$913 \times 500 \times 8 \times 250 = ¥913 \text{ million} \qquad \text{......(2.51)}$$

If the process capability index (the tolerance range divided by six times the standard deviation) could be brought to 1 at an annual cost of ¥500 million, the quality loss L would be:

$$L = 266.7 \times \left(\frac{2.8}{6}\right)^2 = ¥58.1 \qquad \text{......(2.52)}$$

The increased cost would be:

$$\frac{500000000}{500 \times 8 \times 250} = ¥500 \qquad \text{......(2.53)}$$

This time the quality improvement is $1,224 - 58.1 = ¥1,165.9$. Balanced against a cost increase of ¥500, the gain is:

$$1,165.9 - 500 = ¥665.9 \qquad \text{......(2.54)}$$

This is less than the gain in equation 2.50. The above figures can be tabulated as follows.

Table 2.2 Production Cost and Quality Level

Method	Cost (¥)	Variance σ^2	$L = 266.75\sigma^2$ (¥)	Total loss
Present	800	4.59	1224	2024
Proposed (1)	980	0.49	131	1111
Proposed (2)	1300	0.218	58	1358

From Table 2.2, the total loss is least with proposed method (1). This is accordingly the optimum solution.

Destructive inspection cannot be used to improve quality by screening. The only choice is off-line quality control, as in Table 2.2. Off-line quality control is employed for the twin purposes of quality improvement and cost improvement. When the method that minimizes the sum of quality and cost is known, the procedure is not to set the tolerances in relation to the quality level but to set them according to the production (in-factory) cost.

If the present method is replaced by proposed method (1), the cost rises by ¥180 from ¥500 to ¥680. With a value of ¥680 for the parameter A in the tolerance formula, the tolerance Δ becomes:

$$\Delta = \sqrt{\frac{680}{60000}} \times 15 = 1.6 \qquad\qquad ......(2.55)$$

This is only slightly larger than the tolerance of 1.4 given in equation 2.45. In any case, it is important to realize that the tolerance is affected only slightly by the process capability.

When a company orders parts from an affiliate or an outside company, if there is no easy way to rework defectives, the specification tolerance Δ of the purchased parts can be determined using the purchase price. When it is wrong to use the purchase price there is a reworking method that involves a much lower loss than the purchase price. Since reworking is out of the question in cases of destructive inspection, however, it is reasonable to use the purchase price as the value of A in setting the tolerance.

Quality Assessment and Inspection

Suppose the dimension specification of a certain part is:

$$m \pm 5\,\mu m \qquad\qquad ......(2.56)$$

If the loss A caused by being off specification is ¥600, the coefficient k

in the formula for the loss L due to variability is:

$$k = \frac{600}{5^2} = ¥\,24.0 \qquad\qquad ......(2.57)$$

If the mean squared deviation is σ^2, the loss L due to variability is then:

$$L = 24.0\,\sigma^2 \qquad\qquad ......(2.58)$$

Suppose that the parts being made at present have an average characteristic value of m with a standard deviation one-sixth the tolerance range of 10 microns.   For example, suppose the mean dimension of all the parts made during one year (in many cases a period of about three months is used) is m, and the standard deviation σ is:

$$\sigma = \frac{10}{6}\ (\mu\mathrm{m}) \qquad\qquad ......(2.59)$$

The loss L due to variability is then:

$$L = 24.0 \times \left(\frac{10}{6}\right)^2 = ¥\,66.7 \qquad\qquad .....(2.60)$$

One firm is engaged in a campaign to bring its process capability to one- eighth the tolerance range.   This type of campaign calls for engineering measures, but if the square root of the mean squared deviation can be reduced from one-sixth to one-eighth without increasing costs, the gain will be as calculated below. The loss in the latter case is:

$$L = 24.0 \times \left(\frac{10}{8}\right)^2 = ¥\,37.5 \qquad\qquad .....(2.61)$$

Subtracting this from the figure in equation 2.60, we see that quality is improved by ¥ 29.2/unit, or approximately a factor of two. If 500,000 units are produced per month, the monthly quality saving amounts to ¥ 14,600,000. If it costs ¥ 10 per unit to get the standard deviation down to one-eighth of the tolerance range, that is still an improvement of ¥ 19.2/unit.

It is extremely important, however, not to change the specifications in order to improve the process capability. The specifications are for inspection purposes. Process capability should be improved by quality control engineering -- by altering the production technology or the adjustment limits, not by tightening the pass/fail specifications. Inspection should be a method of finding defective or substandard units for removal or reworking, and it cannot have any effect on units that are within specification. Process capability should be improved by improving production engineering and control methods.

Equations 2.60 and 2.61 give the loss when the products are shipped without inspection. If 100% inspection is performed, the value of the variance in equation 2.60 will generally be a little smaller. If the characteristic value is normally distributed, for example, from case 6 in Table 2.3, the proportion of parts outside the specification limits will be 0.27%. If the inspection cost B of 100% inspection is ¥ 3/unit, the total loss when 100% inspection is performed is L, and the variance of the good units by themselves, after the defectives have been removed, is σ^2 out, then:

$$L = \text{inspection cost per unit} + \text{loss due to defectives found}$$
$$\times \text{ fraction defective} + k\sigma^2{}_{out}$$

$$= 3.0 + 600 \times 0.0027 + 24.0 \times 0.986^2 \times \left(\frac{10}{6}\right)^2$$

$$= 3.0 + 1.6 + 64.8$$

$$= ¥\, 69.4 \qquad\qquad\qquad ......(2.62)$$

From the above equation, it is clear that 100% inspection is almost worthless as far as quality improvement is concerned since the percentage of defectives is only 0.27%. The quality improvement is only $66.7 - 64.8 = ¥\,1.9$. Of course, it is a different matter if the purpose of 100% inspection is to screen out very serious defectives. What we are saying is that screening out the tails of the normal distribution does not have much effect.

If the distribution is normal and the standard deviation is one-fourth the tolerance range, the loss L if the units are shipped without inspection is:

$$L = 24.0 \times \left(\frac{10}{4}\right)^2 = ¥\, 150 \qquad\qquad ......(2.63)$$

If 100% inspection is performed before shipment at a cost of ¥ 3/unit, then the percentage of defectives p is 4.55% and the variance of the good units is 0.880^2 of what it was before. From case 4 in Table 2.3:

$$L = 3.0 + 600 \times 0.455 + 24.0 \times 0.880^2 \times \left(\frac{10}{4}\right)^2$$

$$= 3.0 + 27.3 + 116.2$$

$$= ¥\, 146.5 \qquad\qquad\qquad ......(2.64)$$

If monthly production is 200,000 units, this improvement of ¥ 3.5/unit amounts to only ¥ 700,000 per month. Some industrialists from developing countries have asked why their products tend to fail during use after shipment even though they are leasing technology from a leading company in an advanced industrial country they produce according to the specifications set by that company, they use only materials and parts that meet specifications, and they screen out defectives by 100% inspection.

The author's standard answer is that just because products pass inspection does not mean that they are good. Suppose the standard deviation is only half of the tolerance. If the distribution is normal, 31. 7% of the units will fail to meet specifications. If you remove all the defectives by 100% inspection, the standard deviation of the remaining good units (assuming the mean matches the mid-value m of the specification limits) will fall only to 0.539 of the standard deviation before inspection. The loss due to variability will therefore be:

$$L = 24.0 \times (0.539 \times \frac{10}{2})^2$$

$$= ¥ 174.3 \qquad\qquad\qquad ......(2.65)$$

This is not only worse than the loss in equation 2.60; it is even worse quality than the case in equation 2.63 in which the goods are shipped with no inspection, with an admixture of 4.55% defectives. The need is to improve the process capability, not to eliminate defectives by inspection.

Table 2.3 gives values of the loss function L for a variety of cases. A single asterisk (*) means that a normal distribution is assumed, and double asterisks (**) indicate a uniform distribution. When there are no asterisks, the value holds regardless of the distribution. In case 1, the process capability is extremely bad. Even if the defectives are screened out by 100% inspection, the variation in the shipped units causes a loss of ¥ 174.3 (case 2), which is worse than case 3. In case 3 the process capability is still somewhat inadequate, with 4.55% defectives being produced, and the goods are shipped without inspection.

The process capability requirement often calls for goods to be produced with a standard deviation one-sixth the tolerance range. Surprisingly, even without inspection, this leads to only half the loss of case 4. The loss L in Table 2.3 is only the loss due to variability in the shipped products; it does not include the cost of inspection or the loss from defectives discovered in the inspection. If these factors were included, the loss when inspection is performed would be greater.

Table 2.3. Characteristic Value Distribution and Loss

Specification limits m $\pm$ 5 (μm); Loss due to defective unit A $=$ ¥ 600; Loss function $L = 24.0\,\sigma^2$ (¥)

Case	Mean value	Standard deviation $s \cdot d \cdot$	Inspection	Variance σ^2_{out}		Percent found defective (%)	Loss L (¥)	Percent shipped defective(%)
(1)	m	10/2	no	$\left(\frac{10}{2}\right)^2$		0.00	600.0	31.73 *
(2)	m	10/2	yes	$0.539^2 \times \left(\frac{10}{2}\right)^2$	*	31.73 *	174.3 *	0.0 *
(3)	m	10/4	no	$\left(\frac{10}{4}\right)^2$		0.00	150.0	4.55 *
(4)	m	10/4	yes	$0.880^2 \times \left(\frac{10}{4}\right)^2$	*	4.55 *	116.2 *	0.00 *
(5)	m	10/6	no	$\left(\frac{10}{6}\right)^2$		0.00	66.7	0.27 *
(6)	m	10/6	yes	$0.986^2 \times \left(\frac{10}{6}\right)^2$	*	0.27 *	64.8 *	0.00 *
(7)	m	10/8	no	$\left(\frac{10}{8}\right)^2$		0.00	37.5	0.01 *
(8)	m	10/16	no	$\left(\frac{10}{16}\right)^2$		0.00	9.4	0.00 *
(9)	m	$10/\sqrt{12}$	no	$\left(\frac{10}{\sqrt{12}}\right)^2$		0.00	200.0	0.00 **
(10)	$m-2.5$	10/6	no	$2.5^2 + \left(\frac{10}{6}\right)^2$		0.00	216.7	6.68 *
(11)	$m-2.5$	10/12	no	$2.5^2 + \left(\frac{10}{12}\right)^2$		0.00	166.7	0.14 *
(12)	$m-2.5$	10/16	no	$2.5^2 + \left(\frac{10}{16}\right)^2$		0.00	159.4	0.00 *
(13)	$m-2.5$	0	no	2.5^2		0.00	150.0	0.00
(14)	$m-5.0$	0	no	5.0^2		0.00	600.0	0.00

Legend: * = normal distribution
** = uniform distribution
No asterisk = any distribution

Cases 10 to 14 give the loss when the average value deviates from the target value m. The variance (mean square error) increases by the square of the deviation. In case 11, the standard deviation is only 1/12 of the tolerance range, but the average value deviates by 2.5 microns from the target. Although the percent defective is only half of the percent in case 5, the loss due to variability is ¥ 100 larger: ¥ 166.7 as compared to ¥ 66.7. In case 12, the standard deviation is 1/16 of the tolerance range and again the average value is 2.5 microns off target. Despite the complete absence of defectives, the loss is greater than in case 3, and ¥ 92.7 greater than in case 5.

It follows that in production departments, zero defectives does not mean perfect quality. If the characteristic value is a measurable quantity, unending efforts should be made to reduce variability to zero. The tolerances given on drawings are inspection standards, and are not for production control. Although it suffered from unclear economics, the purpose of the Shewhart control-chart method was to assure quality by a control approach instead of a standards-and- inspection approach. Of course this approach is contradicted if the cost of control is greater than the profit gained by reducing variability. The duty of the production department is to reduce variability at a cost well below the profit gained.

In advanced industrial countries, 100% inspection is rarely used. Most characteristic values are distributed around the mid-value m with a standard deviation of one-sixth the tolerance range or less. In the developing countries, 100% inspection is fairly common because of inadequate process capability. This is what leads to the difference in quality. The quality level is not the same as the percentage of defectives; it is a measure of the size of the loss. For a measurable value with specification limits on both sides, it is proportional to the variance. It is important to bear Table 2.3 in mind in selecting or controlling a production process or selecting and evaluating subcontractors.

PROBLEMS

1. The specification limit for NO_x in exhaust gas is 0.48 g/km in driving mode 10 (1978 specification). The cost A of reworking when the specification limit is exceeded is ¥ 12,000, and the recall cost A_0 when the environmental standards are not met is ¥ 50,000. Find the factory shipment specification limit.

2. The operating range (consumer's tolerance range) of the output voltage of a certain circuit is ± 10 V. The social cost A_0 of repair, etc. when the circuit fails is ¥ 6,000. The loss (rework cost) A when a unit is found defective at the factory is ¥ 50. Find the shipping tolerance.

3. A thickness specification is 760 ± 20 (μm), and the loss A when a unit is out of specification is ¥ 300. A sample of 10 units exhibited the following deviations from the target value:

   5, 1, 2, 3, 4, 8, 5, 4, 3, 6

   Construct an analysis-of-variance table for the general mean m

and error e, and compare the losses of continuing as at present and adjusting the mean to the target value.

DISCUSSION

Definition of the Loss Function

S (Student): In equation 2.24, shouldn't the individual loss functions be calculated, then averaged?

G (Genichi): You could calculate individually, and the glass would then fit the building when it was built, but the window frames might later get warped out of shape due to an earthquake. The glass manufacturer does not know where his products will be used, or how long they will be used. Let y be the width of the pane of glass, p_i the proportion that are used at the i-th site, and $L_i(t, y)$ the loss after t years. If the design life is T years, the expected loss $L(y)$ is defined by the equation:

$$L(y) = \sum_{i=1}^{n} p_i \int_{0}^{T} L_i (t, y) \, dt \qquad \qquad ......(\text{D-2.1})$$

S: This seems to be the average of the loss function if the product is used for T years at each site. Presumably the calculations include the possibility that the frame supporting the glass may be deformed, causing the glass to break or fall out, with injury to people below.

G: When a pane of glass is measured to have a width y, the site where it will be used is unknown, so you have to consider the probability of its being used at all the sites. But it would be impossible to determine those probabilities and find the loss function $L_i(t, y)$ at each site. That is why the loss function is determined by expanding $L(y)$ around the target value m. It is safe to assume that $L(y)$ is a continuous function of y.

S: Even if the functions for the individual sites are noncontinuous, when they are averaged probabilistically under all the different conditions, the result will be approximately continuous. It makes sense that $L(y)$ should have a Taylor expansion around the target value m. The problem is in determining the coefficient k. You're determining k from the consumer tolerance range and the loss at those points, but the reasoning is a little hard to follow.

G: The value of k should be determined by calculating the loss function at a point as close as possible to the target m. The consumer tolerance limits would seem to be the only points available.

S: Calculating k from the consumer tolerance range and the social loss caused by being outside that range does seem to be the easiest way.

G: If the loss at a point closer to the target value were known, it would be better to determine k at that point, but ....

S: The *LD*50 point does seem the easiest to understand.

Harmful Components and Other Cases in Which Less is Better

S: If we are dealing with the side effects of a drug and the *LD*50 point is 100 mg, how do we calculate the cost of the patient's dying?

G: I guess we do it by considering per-capita income. At present, annual per-capita income is about ¥ 1,300,000 per year, so multiply it by the average life expectancy. If the mean for male and female life expectancies is 75 years, the loss A_0 caused by going outside the tolerance limit is:

$$A_0 = ¥\,1,300,000 \times 75 = ¥\,97,500,000 \qquad ......(D\text{-}2.2)$$

It follows that the coefficient k is:

$$k = \frac{A_0}{100^2} = \frac{97500000}{10000} = ¥\,9,750 \qquad ......(D\text{-}2.3)$$

S: To determine the upper limit Δ on the harmful component, we need to know the loss when the drug fails to meet the specification. Since the drug has to be discarded when it fails to meet specification, suppose that the loss A is ¥ 300. The tolerance Δ is found as follows from the formula for the one- sided case.

$$\Delta = \sqrt{\frac{A}{A_0}} \times \Delta_0 = \sqrt{\frac{300}{97500000}} \times 100 = 0.18 \ (\text{mg}) \qquad ......(D\text{-}2.4)$$

Accordingly, the manufacturing tolerance should be 1.8 thousandths of the *LD*50 point of 100 mg, which is the consumers tolerance limit.

G: Right.That's a very strict tolerance, but it's what our theory gives. We're assuming that the harmful component can be removed by present technology. If it can't, the tolerance can be found from the formula:

$$\frac{300}{\text{Fraction conforming}} = 9,750\,\Delta^2 \qquad ......(D\text{-}2.5)$$

If we solve this we get:

$$\Delta = \sqrt{\frac{300}{9{,}750 \times \text{fraction conforming}}}$$

$$= 0.175 \times \frac{1}{\sqrt{\text{fraction conforming}}} \qquad ......(D\text{-}2.6)$$

S: This last equation is not easy to solve, because the fraction conforming to the tolerance depends on Δ.

G: The only way to solve it is by successive approximations. You can refresh your memory with the Section entitled Inadequate Process Capability.

S: What about the case of automobile exhaust? Here the harmful components do not come from individual cars.

G: Like pollution problems in general, this is a difficult question. The $LD50$ point is the point at which half the population of Japan dies. In particularly bad locations, the CO concentration is about 1.0 ppm. Let's suppose that half of the people would die at a CO concentration of 0.3%. Now we have to find what concentration of CO in exhaust gas it would take to raise the CO concentration throughout Japan to that $LD50$. Let's suppose that in a certain driving mode average of CO is 0.4 g/km, and that if the concentration reached 400 times the present level, half the population of Japan would die.

S: Under those assumptions, the $LD50$ point, the consumer tolerance limit, is:

$$0.4\,\text{g} \times 400 = 160\,\text{g} \qquad ......(D\text{-}2.7)$$

The loss depends on how you value human life, but perhaps we can use the figure of ¥ 97.5 million for A_0.

G: Japan's human population and automobile population are different. The value of A_0 is as follows.

$$A_0 = \frac{\text{human population}}{\text{automobile population}} \times ¥\,97.5\,\text{million}$$

$$= \frac{110{,}000{,}000}{30{,}000{,}000} \times 97.5 = ¥\,357.5\,\text{million} \qquad ......(D\text{-}2.8)$$

S: In that case the loss function is given by the following formula:

$$L = \frac{A_0}{\Delta_0{}^2}\,y^2 = \frac{357{,}500{,}000}{160^2}\,y^2 = 13{,}965y^2 \qquad ......(D\text{-}2.9)$$

If we assume that a unit that exceeds the tolerance can be reworked at a cost of ¥ 12,000 with present technology, then:

$$\varDelta = \sqrt{\frac{12,000}{357,500,000}} \times 1.60 = 0.9 \ \mathrm{g/km} \qquad \text{......(D-2.10)}$$

This is a little less stringent than the present regulations.

G: I think the value of $\dot{A}_0$ is appropriate, but I feel that a slightly smaller value should be used for $\varDelta_0$. The point is that this type of estimation gives roughly the correct answer. The purpose of this theory is to indicate what should be studied in setting tolerances. You can't expect progress to come from political methods such as collecting a group of consumer and manufacturer representatives, having them voice their opinions, then deciding by majority vote.

S: Advances in technology will reduce the cost A, in which case the specifications will become more stringent. If the number of cars increases, the specifications will again become more stringent, at the rate of the square root of the number of cars. This theory seemed extremely rough, but I see that it gives quite a bit of information, at least in the form of approximations.

G: In many cases, it is not only a first-approximation method of setting tolerances; it is the only available method. In the example we're discussing, there are many different locations inhabited by people with many individual differences. This loss function can be thought of as the result of averaging all those conditions.

Loss Function

S: It is difficult to determine the target value of the thickness of sheet steel for automobile bodies. Reducing the thickness improves gasoline mileage, but then the service life is shorter and the danger in a collision is greater. It's hard to see how to set the optimum target value.

G: I'd like to save that question for a later chapter. Let's assume the optimum value is given and consider the problem of the tolerance.

S: Let's take 760 μm as the target value. When it comes to the tolerance, if we want to find the loss caused by variability of products on the market, we need to know the behavior of the loss function when the target value is changed.

G: If we're dealing with sheet steel for automobile bodies, the user most sensitive to loss due to variability is likely to be the factory that presses the sheets into body shapes. The press factory sets up its press equipment and molds on the assumption that it will be getting sheets 760 μm thick. To set the tolerance, you would have to find out at what deviation from 760 μm the dimensions after pressing would deviate from their nominal values so much as to require reworking.

What should the sheet steel manufacturer do if the variability of the product he ships does not meet the user's tolerance requirements? This is a real problem. The sheet steel manufacturers have their own tolerance specifications, which frequently do not meet the requirements of the body manufacturers.

Tolerances are set according to economics. In setting tolerances, you have to think about what will be done with the nonconforming products, both at the manufacturer's end and the user's end. One company which stamps sheet steel used in industrial refrigerators measures the thickness of the stock it purchases, divides it into three groups -- a too-thin group, a just- right group, and a too-thick group -- and stamps each group separately. That way it is able to use even the nonconforming sheets by altering the press setup. The cost of processing nonconforming sheets is therefore:

$$A = \frac{\text{additional cost of stamping sheets in three groups}}{\text{number of nonconforming (too-thin or too-thick) sheets}}$$

If the value of A is smaller than the manufacturer's loss from rejecting nonconforming sheets, the smaller A should be used to set the tolerances.

S: When the manufacturer produces a bad sheet, he can either scrap it or sell it as second-rate goods. The loss here is greater than having the buyer sort the sheets by thickness and work them accordingly. In this case the consumer's value of A should be used.

G: That's true, but if the thickness of the sheet can be controlled without raising the cost, then the buyer needn't sort the sheets and stamp them under different conditions. As matters stand it would be correct to use the consumer's value of A, but it would be better if the manufacturer could apply off-line quality control to the sheet thickness.

S: The correct way to set the tolerances is the way that makes them the most stringent.

Destructive Testing and Tolerances

S: Since destructive testing precludes 100% inspection, I thought the tolerances should be set according to the process capability, but the result in the text indicates that they should be adjusted to the user's requirements. I find this point hard to understand. Can you explain it a little more?

G: The values of the parameters A_0 and Δ_0 are determined at the user's or consumer's end. The question is the value of A, the loss incurred by the manufacturer of a rejected product. In the case of destructive testing, the defectives cannot all be screened out by inspection. If some lots have many defectives and some have only a few, it would be possible to ship only the lots with few defectives, but the real problem is one of process control. If the process were under control, this situation would not occur.

S: If the process control problem comes first, we have to assume that it has been solved and all lots contain about the same percent of defectives. But I'd also like a simple discussion of the case when there are differences between lots.

G: If there are no differences between lots, then the lots are either all accepted or all rejected. If they are all rejected, the user does not have any material to work with.

S: In other words, either you pass all lots or else you stop production. In that case tolerances and specifications don't seem to have any meaning.

G: I wouldn't say that. If no tolerances are given, you'll always be worried about how badly the process might be doing. Also, there will be no way to make the manufacturer aware of his quality level or put pressure on him to improve.

S: If you're going to pass all lots anyway, there's no harm in setting a tolerance of $\pm\sqrt{A/A_0} \times \Delta_0$, Where A is the in-factory cost or the shipped cost. But it goes against common sense to set a tight tolerance so that there is a high percentage of defectives and then purchase the lot anyway. Shouldn't the tolerance be a little wider?

G: The control limits used in the manufacturing process should be wide, but the value I think the purchaser should use to evaluate quality is:

$$L = k\sigma^2$$

If the quality is hardness, for example, he could measure the hardness of one or two pieces from each lot to calculate the mean squared deviation σ^2 from the target value every month, or every three months, and then use the formula $L = k\sigma^2$ as a test of quality. This is for evaluation purposes; it is not an acceptance inspection.

S: The only time destructive testing should be used to decide whether to accept or reject a delivered lot is when the manufacturer is not performing quality control properly. If the factory is operating under stable conditions, both acceptance and shipping inspections are pointless. But what if the process is unstable?

G: Unstable means that there are major differences in quality between lots. If there are data that will give the value of the expression for quality level L for each lot, I think you can reject a lot if the loss it causes is much larger than the average loss value in a lot produced under normal conditions.

S: If the loss per unit under normal conditions is ¥ 1,224, and the in-factory cost is ¥ 500, then a lot should be rejected when the value of the loss function is ¥ 1,724 or greater.

G: That's right. The loss the manufacturer takes from a reject is the cost of remaking it, so the in-factory cost seems appropriate.

S: About what size of sample is needed to calculate the loss?

G: It is the supplier's job to determine whether individual lots pass or fail. This problem is not really a sampling problem, and it's best treated as a quality control problem. But let's save that for Chapter 5. See reference (5).

3

DETERMINING TOLERANCES

Higher- and Lower-level Characteristics

For a manufacturer of parts or materials, the specification limits are usually set by the firm buying the products. For a manufacturer of finished products, the product specifications are often set by the planning department, as noted in the previous chapter. In the design process, the purpose of parameter design and tolerance design is to set specifications for the contributing characteristics that affect the above objective characteristics. The question in parameter design is how to specify lower-level characteristics. Leaving the determination of their nominal values for a later chapter, here we will assume that the nominal values have already been decided and consider how to determine the tolerances.

When the shipping specifications for a product characteristic are given, the characteristic is a higher-level characteristic in relation to the characteristic values of the component subsystems and parts. Similarly, the subsystem characteristic values are affected by the characteristic values of parts and materials, so they are higher-level in relation to those characteristic values. If the characteristic values of a firm's shipped products affect the characteristic values after processing by the user firm, then the characteristic values of the user firm are higher-level characteristic values.

Hardness and Thickness of Sheet Steel

A firm produces products stamped from sheet steel. If the product taken from the stamp has the wrong dimension, it must be reworked at a cost A_0 of ¥ 1,200. The specification for the dimension in question is m ± 300 μm, and this dimension is affected by the hardness and thickness of the sheet steel. If the hardness varies by one unit on the Rockwell hardness scale, the dimension varies by 180 μm. If the thickness of the sheet varies by 1 μm, the product dimension varies by 6 μm. We shall determine the hardness and thickness tolerances, assuming that sheets that do not meet either specification will be scrapped at a loss of ¥ 300 per stamped product.

Problems in setting tolerances for contributing (lower-level) characteristics are solved by writing the formula for the loss function of the higher-level characteristic and converting it to a formula for the contributing characteristic. If the specification for the higher-level characteristic is $m_0 \pm \Delta_0$ and the loss caused by not meeting the specification is A_0, then the loss function L is given by the formula:

$$L = \frac{A_0}{\Delta_0^2} (y - m_0)^2 \qquad \text{......(3.1)}$$

Let x be the lower-level characteristic and β the effect on the higher-level characteristic when x varies by a unit amount. The right side of the above equation is then:

$$\frac{A_0}{\Delta_0^2} [\beta (x - m)]^2 \qquad \text{......(3.2)}$$

Here m is the nominal value of the lower-level characteristic. Inserting this value into the right side of equation 3.1 and substituting the loss A when the lower-level characteristic fails to meet its specification for the loss L on the left side of 3.1, we have:

$$A = \frac{A_0}{\Delta_0^2} [\beta (x - m)]^2 \qquad \text{......(3.3)}$$

Solving this for $\Delta = (x - m)$, we see that the tolerance for the lower-level characteristic x is given by the formula:

$$\Delta = \sqrt{\frac{A}{A_0}} \times \frac{\Delta_0}{\beta} \qquad \text{......(3.4)}$$

where,

A_0 = loss when higher-level characteristic (objective characteristic) does not meet specification

Δ_0 = tolerance of higher-level characteristic

A = loss when lower-level characteristic does not meet specification

β = effect on higher-level characteristic of unit variation in lower-level characteristic

A_0 and A are the losses when the departure from specification is discovered at the factory where the product is made or the firm to which it is shipped. They do not include inspection cost. A_0 is the loss the manufacturer suffers due to reworking or scrapping the product when it fails to meet the manufacturer's own specifications, or the loss he suffers if it causes the user firm's higher-level characteristic to be out of specification.

In this problem, for the hardness characteristic, A_0 is the loss when the dimension of the stamped product is out of specification. Since:

$$A_0 = ¥\,1,200$$
$$A = ¥\,300$$
$$\Delta_0 = 300\,\mu m$$
$$\beta = 180\,\mu m$$

the hardness tolerance Δ is:

$$\Delta = \sqrt{\frac{300}{1200}} \times \frac{300}{180} = 0.83 \quad (H_R) \qquad\qquad ......(3.5)$$

The shipping specifications for hardness are accordingly $m \pm 0.83$ (H_R).

For thickness of the steel sheet, A_0, A, and Δ_0 are the same as before, and $\beta = 6\,\mu m$. Formula 3.4 can be used to calculate the thickness tolerance as follows.

$$\Delta = \sqrt{\frac{300}{1200}} \times \frac{300}{6} = 25.0 \quad (\mu m) \qquad\qquad ......(3.6)$$

Accordingly, the thickness tolerance of the shipped product should be $\pm 25.0\,\mu m$. We are assuming, of course, that the mean thickness value is the value such that the dimension will have the target value after the sheet is stamped. The tolerances in this problem are being set by the firm that does the stamp work. If the sheet steel manufacturer does not make products that are within these tolerances, the loss A from defective sheets will be different.

If defectives have to be scrapped, the value of the parameter A is the cost of making a replacement that meets the specification. If the manufacturing cost is ¥ 300 and the nondefective yield is only 50%, then there is only a 50% probability that any replacement manufactured will meet the specification. The value of A must therefore be doubled to ¥ 600. If the percentage of nondefectives is

80%, since there is only a small difference between ¥ 300 and the true loss (300/0.8 = ¥ 375), the solution Δ = 25.0 μm obtained from A = ¥ 300 can be used. If the percentage of nondefectives is less than 80%, the tolerance must be found by solving equation 3.7, in which $q(\Delta)$ is the fraction nondefective when the tolerance is Δ.

$$\Delta = \sqrt{\frac{A}{A_0 q\,(\Delta)}} \times \frac{\Delta_0}{\beta} \qquad\qquad ......(3.7)$$

In most cases, however, it is correct to use equation 3.4, not equation 3.7, to determine the tolerance, because it is the task of production engineering to improve the process so as to meet the tolerance Δ.

Deterioration Characteristics

Characteristics are said to deteriorate if they change over time. Parts may gradually wear out, for instance, or the resistance values of the resistors in an electrical circuit may gradually increase. In many cases inspection is difficult, and special techniques are needed to set specifications for these characteristics.

If the number of seconds a watch gains or loses a month can be predicted in a few seconds, this accuracy is not a deterioration characteristic, but we shall consider it in the same category as a deterioration characteristic. Similarly, if a characteristic varies with temperature, it may be necessary to set a tolerance for the degree of variation, although this is not called a deterioration characteristic but is called a temperature characteristic or temperature coefficient.

Since temperature characteristics and the watch accuracy can be inspected, tolerances are frequently set on the assumption of inspectability. When a characteristic deteriorates with the passage of time, the effect of the deterioration varies depending on how much time is allowed. It is thus necessary to decide the design life -- be it 10 or 20 years -- in advance. The same goes for environmental effects and the watch accuracy.

Tolerances for Deterioration Characteristics

Let us consider the case in which a characteristic degrades at a nearly steady rate. If the design life of the product is T years, and the characteristic degrades by βT in T years, then β is the amount of deterioration in one year. In the case of a watch, β is the amount of time the watch gains or loses a month.

If the characteristic value of the product starts out at the nominal value m and changes by β per year, at the end of T years it will deviate from the nominal by βT. As it gradually drifts away from the nominal value, the mean squared deviation σ^2 is given by the following integral.

$$\sigma^2 = \frac{1}{T} \int_0^T (\beta t)^2 \; dt = \frac{T^2}{3} \beta^2 \qquad\qquad ......(3.8)$$

As before, we shall refer to the mean squared deviation as the variance. In many cases deterioration can be corrected; for example, a watch that runs fast or slow can be adjusted by changing the capacitance of a capacitor.

Suppose that the specified accuracy of a watch is ± 10 seconds per month. If the accuracy is within this specification and the user sets the watch once a month, it will always be accurate to within ± 10 seconds. If the watch does not meet this specification, the consumer will have to set it more often. Let us put the cost of adjusting the watch at ¥ 120. We shall also assume that the watch gains or loses time at a nearly constant rate of β seconds per month.

The consumer buys the watch on the strength of its ± 10 seconds/month accuracy specification. Suppose that the cost of checking and adjusting the accuracy of the watch at the end of the manufacturing process is ¥ 600. If the design life of the watch is 10 years (120 months), the loss L due to deviation at the rate of β per month will be:

$$L = k\sigma^2 = k \times \frac{1}{1} \int_0^1 (\beta t)^2 \; dt = \frac{120 \times 120 \, \text{month}}{10^2} \times \frac{1}{3} \times \beta^2 \qquad ......(3.9)$$

Substituting the adjustment cost of ¥ 600 on the left side, we have:

$$¥\, 600 = 144 \times \frac{\beta^2}{3}$$

$$\beta = 3.5 \text{ seconds/month} \qquad\qquad ......(3.10)$$

The accuracy specification for shipment from the factory should be ± 3.5 seconds. This calculation can be put into a formula as follows. If a deterioration coefficient of β can be corrected to meet the specification at a cost of A, the design life is T, and a loss of A_0 occurs when the characteristic value is Δ_0'. The tolerance Δ of β can be found by solving the following equation for β.

$$A = \frac{A_0}{\Delta_0^2} \times \frac{\beta^2}{3} \times T^2$$

The tolerance Δ is:

$$\Delta = \sqrt{\frac{3A}{A_0} \times \frac{\Delta_0}{T}} \qquad\qquad ......(3.11)$$

where

A = cost of adjusting (repairing) the deterioration coefficient
Δ_0 = tolerance limit of characteristic value (on the market)
A_0 = social loss when characteristic value is outside the above limit
T = design life or calibration period

For the accuracy of the watch in the above example, $A = ¥600$, $\Delta_0 = 10$ seconds, $A_0 = 120 \times ¥120$, and $T =$ one month. Substituting these values into the preceding formula,

$$\Delta = \sqrt{\frac{3 \times 600}{14400}} \times \frac{10}{1} = 3.5 \ (\text{sec}) \qquad\qquad ......(3.12)$$

This is the same result as obtained in 3.10.

If it is difficult to control the accuracy to within this range, one should find the percentage within the specification after adjustment to the best possible accuracy. If 70% are within the specification, the value of Δ should be recalculated as follows:

$$\Delta = \sqrt{\frac{3A}{qA_0} \times \frac{\Delta_0}{T}} = \sqrt{\frac{3 \times 600}{0.7 \times 14400}} \times \frac{10}{1} = 4.2 \ (\text{sec}) \qquad\qquad ......(3.13)$$

This is the second approximation to the accuracy. If 85% are within specification when the accuracy specification is ± 4.2 seconds, Δ can be found again to obtain a third approximation:

$$\Delta = \sqrt{\frac{3 \times 600}{0.85 \times 14400}} \times \frac{10}{1} = 3.8 \ (\text{sec}) \qquad\qquad ......(3.14)$$

The third approximation is 3.8 seconds.

This method of determining the tolerance in consideration of the process capability is for use in setting internal standards for products sold to the general public. If the product is sold to a user company and that company sets a tolerance, the process should be designed to meet the tolerance. As long as any competitor meets the tolerance, it is essential to improve your own firm's process capability. The above calculations are for when your firm is responsible to society at large and the process capability cannot be improved.

When Measurement and Adjustment are Impossible;

The accuracy of a watch or sensitivity of an instrument can be measured and calibrated, but there are many products for which this is not possible. Examples include the deterioration   of electronic components. We shall consider the case of the components of the power circuit of a television set.

Television power circuits convert 100-V alternating current to 115-V direct current. In theory, the 115 V nominal output voltage is selected in relation to the total system design of the television, although in Japan the American value was simply adopted without any independent study. Whether the value of 115 V is arrived at by a rational process or not, however, once it is decided on, the tuning circuits and the picture tube are designed on the assumption that the power supply will deliver 115 VDC.

If the output voltage of the power circuit deviates from 115 V, there will accordingly be a loss. If the output value is y, then as in the preceding chapter, the loss function $L(y)$ due to deviation from the 115-V target is given as follows.

$$L(y) \fallingdotseq k(y - 115)^2 \qquad\qquad ......(3.15)$$

A family buys a television set and uses it for a number of years. Due to deterioration of components and other reasons, the output begins to vary from 115 V. Let's assume it drops below 90 V. The picture is too dim and the contrast too weak to be corrected by the adjustment controls; either the power circuit must be repaired or the television set must be replaced. We shall put the average loss to the consumer A_0 at ¥ 30,000. Similarly, if the output voltage becomes too high, the set gives trouble, circuit components are damaged, the contrast is too strong, and again the set must be repaired or replaced. In general, the consumer's tolerance on the high side will be different from that on the low side, but for simplicity's sake, we will assume here that roughly the same loss occurs on both sides. That is, we are making the following assumptions about the output voltage of the power circuit:

Consumer tolerance range = 115 ± 25 V $\qquad$ ......(3.16)

Average consumer's loss when outside tolerance range
$$A_0 = ¥ 30,000 \qquad ......(3.17)$$

If the tolerance range varies with the consumer, we can take the $LD50$ point as the value at which 50% of the people would repair or replace the set or lodge a complaint. The loss function $L(y)$ is an average obtained by considering the probabilities that the television set will be used under all possible conditions. Although this is a highly complex function in reality, we shall approximate it by a quadratic expression.

When the output voltage value y goes below 90 V or above 140 V, 50% of consumers will lose patience and have the set repaired or replaced at an average loss of ¥ 30,000 apiece, so the value of the loss function at $y = 90$ and $y = 140$ is approximately ¥ 30,000. This statement is justified on the grounds that the reason the consumer has the set repaired or replaced despite the ¥ 30,000 cost is that if the output voltage goes outside the tolerance range, he considers the loss to be ¥ 30,000 or more. We can therefore substitute ¥ 30,000 for the left side of equation 3.15 and 25 for $(y - 115)$ on the right side, obtaining:

$$30,000 = k \times 25^2 \qquad\qquad ......(3.18)$$

Solving this for the coefficient k, we have:

$$k = \frac{30000}{25^2} = ¥\,48.0 \qquad\qquad ......(3.19)$$

The loss function is accordingly approximated by the following expression:

$$L = 48.0(y - 115)^2 \qquad\qquad ......(3.20)$$

In general, the tolerance on the drawings used by the production department that manufactures the power circuit will be entirely different   from the consumer tolerance, because if the production process falters and turns out a unit that does not meet the drawing specifications, it can usually be reworked or scrapped at a much lower loss. Suppose the power circuit can be fixed by replacing one of its resistors if the output voltage does not meet the drawing specification. If a 50-kilohm resistor is usually used, the output voltage can be raised by 0.5 V by lessening the resistance by 1 kilohm. If the output voltage is 112 V when the input is 100 VAC, replacing the 50-kilohm resistor with a 44- kilohm resistor will adjust the output voltage to the target value of 115 V. The cost of the resistor is about ¥ 10, and the replacement can be made in two or three man-minutes, so the adjustment cost is only about ¥ 100.

An output voltage of 112 V is still within the consumer's tolerance range. If the manufacturer decides to save ¥ 100 by shipping the unit without adjusting it, he will cause the consumer a loss given by formula 3.20 of:

$$L = 48.0(112 - 115)^2 = ¥\,432 \qquad\qquad ......(3.21)$$

To inflict a loss of ¥ 432 on the consumer in order to save yourself ¥ 100 is worse than criminal. When a criminal steals ¥ 10,000, he gains ¥ 10,000 while his victim loses ¥ 10,000, so society as a whole comes out even. Since it is wrong to ship a power circuit with an output voltage of 112 V, such a circuit should not be considered as meeting specification.

The manufacturing specification should be determined by substituting the loss $A = ¥ 100$ when a unit is rejected at the manufacturing stage on the left side of the equation for the loss function and solving for y.

$$100 = 48.0(y - 115)^2$$

Solving for y gives the following specification:

$$y = 115 \pm \sqrt{\frac{100}{48.0}} = 115 \pm 1.4 \text{ V} \qquad \qquad ......(3.22)$$

Although the consumer tolerance range is 115 ± 25 V, the manufacturing tolerance range should be 115 ± 1.4 V.

Let us consider the meaning of the ¥ 432 loss in equation 3.21 a little further. This is an actual loss suffered by the consumer. The consumer tolerance is ±25 V. Consider a television set in which the output voltage of the power circuit is 95 V. If a house is situated at some distance from the transformer, the AC voltage received at that house may well be about 80 V. If a family living in such a house purchases this television set, the output voltage will be less than 90 V from the very beginning. Even in more fortunate houses, the service life of the set will be shorter than the life of a set with the nominal output voltage of 115 V. The ¥ 432 loss is the average loss suffered by all households in the from of shorter service life and a narrower voltage tolerance at the wall outlet.

Next let us suppose that the nominal value of the resistance of a resistor in the power circuit is m kilohms, and consider what the tolerance should be. The resistance value x has a linear effect on the output voltage in the range in which we are interested. Actually, a linear effect would be unthinkable after adequate parameter design, but since there are other target characteristics, such as current, it is frequently impossible to design the parameters to the optimal extreme values. Let β be the effect on the output voltage y when the resistance value x varies by one kilohm. We shall assume that:

$$\beta = 0.5 \text{ V/kilohm}$$

Accordingly, the tolerance of the resistance value x around nominal value m is given by the formula:

$$m \pm \sqrt{\frac{A}{A_0}} \times \frac{1}{\beta} \times \text{tolerance of higher-level characteristic}$$
......(3.23)

where

A_0 = loss when the power circuit does not meet its specification

A = loss when the resistor is rejected

If A_0 = ¥100 and A = ¥10, then since the tolerance of the higher-level characteristic is 1.4 V,

$$m \pm \sqrt{\frac{10}{100}} \times \frac{1}{0.5} \times 1.4 = m \pm 0.89 \text{ kilohms}$$
......(3.24)

In general, if the tolerance of the next-higher-level target characteristic is Δ_0, the loss when it is outside the tolerance range is A_0, the effect on the higher- level characteristic of a unit change in system component parameter x is β, and the loss when that system component is rejected is A, then the tolerance of the system component parameter is given by the formula:

$$\pm\sqrt{\frac{A}{A_0}} \times \frac{\Delta_0}{\beta}$$
......(3.25)

A tolerance for a deterioration coefficient can be determined as follows. Let b be the per-year deterioration coefficient, T = 10 years be the design life, and m be the nominal value of the initial characteristic. If the deterioration is linear, the mean square deviation from the nominal value over the ten years is given by the formula:

$$\sigma^2 = \frac{1}{T}\int_0^T (bt)^2\, dt = \frac{T^2}{3}b^2 = \frac{100}{3}b^2$$
......(3.26)

If the above value of σ^2 is substituted for $(y - 115)^2$ on the right side of the loss function L in 3.20, then:

$$L = 48.0\left[\beta^2 \times \frac{100}{3}b^2 \right]$$

Components with a deterioration coefficient outside the specification limits are scrapped. If the loss is A = ¥10, we have the following equation:

$$10 = 48.0 \left[0.5^2 \times \frac{100}{3} b^2 \right]$$

Solving this for b, we find the tolerance of the deterioration b per year to be:

$$| b | = \sqrt{\frac{10}{48} \times \frac{1}{0.5^2} \times \frac{3}{100}} = 0.158 \qquad \qquad ......(3.27)$$

That is, the tolerance of the deterioration coefficient b is 0 ± 0.158.

Such specifications are standards to be used in deciding which characteristic values of a component to accept or reject. They are like setting 60% as the passing score on a test in school. The fact that a student passes with a 60% does not mean that he is a good student. For both finished products and components, the mean squared deviation from the nominal value must be found and substituted into the loss function to determine the quality level, at least in relation to function.

If m is the mean initial characteristic value produced by the manufacturer of the resistor, its standard deviation is one-third the tolerance of 0.89, the nominal value of the deterioration coefficient is 0.0125, its standard deviation is one-third of its tolerance of 0.158, and a is the deviation of the initial value from the nominal value, then the variance o^2 is as follows:

$$o^2 = \text{mean of } a^2 + \text{mean of } 10ab + 100/3 \, (\text{mean } b^2)$$

$$= \left(\frac{0.89}{3} \right)^2 + 0 + \frac{100}{3} \left[0.0125^2 + \left(\frac{0.158}{3} \right)^2 \right]$$

$$= 0.186 \qquad \qquad ......(3.28)$$

From this, the loss L due to variability is:

$$L = 48.0 \times 0.5^2 \times 0.186 = ¥ 2.23 \qquad \qquad ......(3.29)$$

If one million of these resistors are produced per year, the annual loss due to variability is ¥ 2.23 million.

In general, let $m_0 \pm \Delta_0$ be the specification for the higher-level characteristic and let A_0 be the loss when it is out of specification. The tolerance Δ of the initial value of a lower-level characteristic x that affects the higher-level characteristic y at a rate of β per unit change is given by formula 3.30 below. The tolerance Δ^* of the amount of change b per year is given by formula 3.31.

$$\Delta = \sqrt{\frac{A}{A_0}} \times \frac{\Delta_0}{| \beta |} \qquad \qquad ......(3.30)$$

where,

$A_0 =$ loss when the higher-level characteristic value is outside the specification limits

$\Delta_0 =$ tolerance of higher-level characteristic

$A =$ loss when the lower-level (component) characteristic value is outside its specification limits

$\beta =$ variation in the higher-level characteristic per unit variation in the lower-level characteristic value

Solving this for Δ^*, we get:

$$\Delta^* = \sqrt{\frac{3A^*}{A_0}} \times \frac{\Delta_0}{|\beta|\,T}$$

......(3.31)

A_0, Δ_0, and β in equation 3.31 are the same as in 3.30, and A^* and T are:

$A^* =$ loss when the deterioration coefficient of the lower-level (component) characteristic is outside its specification limit

$T =$ design life (in years)

Suppose, for example, that when an illuminance parameter varies by 50 lux, quality problems appear and repair or replacement is necessary at a loss to society of about ¥ 15,000.

$\Delta_0 = 50$ lux
$A_0 = $ ¥ 15,000

Suppose that if the lamp intensity varies by a unit of 1 candela in the manufacturing process, the illuminance changes by 0.8 lux. Let the adjustment cost A if the initial intensity of the lamp is outside the specification limit be ¥ 300, and let the loss A^* due to scrapping it if it departs from the deterioration specification limit be ¥ 3,200. Then:

$\beta = 0.8$ lux/candela
$A = $ ¥ 300
$A^* = $ ¥ 3,200
$T = 20,000$ hours

Using these values, the tolerance Δ of the intensity and the tolerance Δ^* of the deterioration can be found from formulas 3.30 and 3.31 as:

$$\Delta = \sqrt{\frac{A}{A_0}} \times \frac{\Delta_0}{|\beta|} = \sqrt{\frac{300}{15000}} \times \frac{50}{0.8} \fallingdotseq 8.8 \ \ (\text{cd})$$

......(3.32)

$$\Delta^* = \sqrt{\frac{3A^*}{A_0}} \times \frac{\Delta_0}{|\beta|} \times \frac{1}{T} = \sqrt{\frac{3\times3200}{15000}} \times \frac{50}{0.8} \times \frac{1}{20000}$$

$$= 0.0025 \ (cd/h) \qquad\qquad ......(3.33)$$

The tolerance of the initial intensity of the lamp is ± 8.8 cd, and the tolerance of the coefficient of deterioration is 0.0025 cd/h or less. If dirt, for example, reduces the illuminance while the lamp intensity is still satisfactory, then the cleaning interval should be used instead of the design life in these calculations.

PROBLEMS

1. A circuit has a nominal output value (lower limit) in the on state of m mV, with a tolerance of 5 mV. If the circuit is outside its specification limits ($m \pm 5$ mV), it is scrapped at a loss of ¥ 20. When a resistance in the circuit varies by 1 m Ω, the output voltage varies by 0.5 mV. The target value of this resistance is selected to give the target output voltage of m mV. Find the tolerance. The loss from rejecting the resistor is ¥ 5.

2. The accuracy tolerance when a certain watch is shipped is ± 3 seconds/month. Watches that are outside this tolerance are calibrated by adjusting the capacitance of a capacitor at a cost of ¥ 600. Since the frequency of oscillation of the quartz crystal used in the watch varies with temperature, the rate of variation was studied for two types of crystals, A_1 and A_2, at five different temperatures B, with the results shown below. The figures are the variation from the frequency at 20°C expressed in parts per million.

	0 ℃ B_1	10℃ B_2	20℃ B_3	30℃ B_4	40℃ B_5
A_1	30	14	0	−15	−29
A_2	24	13	0	−11	−23

(1) Type A_2 cost ¥ 50 more than A_1. Which type should be used? The standard deviation of the temperature under field conditions is 6°C, and the price of A_1 is ¥ 200.
(2) After determining which crystal to use, find the tolerance of its temperature characteristic.
(Hint) Make an analysis-of-variance table by separating the linear

effects from the remaining terms, using the orthogonal polynomials for both A_1 and A_2.

3. The nominal value for a dimension at a certain location in a product is $300 \, \mu m$. If the dimension wears down by $120 \, \mu m$, trouble will occur in the field, with a loss A_0 of ¥ 20,000. Find the tolerances Δ and Δ^* of the initial value and rate of wear of the dimension of a component that affects the above objective dimension. If either the initial value or rate of wear is out of specification, the component is scrapped at a loss of ¥ 50. When the dimension of the component varies by $1 \, \mu m$, the objective dimension also varies by $1 \, \mu m$. Find the rate-of-wear specification for a design life of 10 years.

If the best obtainable value for the rate of wear of the component dimension is $2.0 \, \mu m$/year, how many times during the 10 years should this component receive maintenance? Assume that preventive maintenance should be performed at the optimum amount of wear. If preventive maintenance costs ¥ 3,000, find the loss (maintenance cost + loss due to variation).

DISCUSSION

The Problem of Allocation of Tolerances

S (Student):I thought that tolerance design was a process of allocating tolerances among lower-level characteristics that affect a higher-level (objective) characteristic so that the higher-level characteristic will meet its tolerance specification, but the method of this chapter assigns tolerances to the lower- level characteristics one by one. As a result, isn't it possible that the higher- level characteristic could be outside its tolerance range even if all the lower- level characteristics are within their tolerances?

G (Genichi):The idea of allocating tolerances among the lower-level characteristics (system parameters) so that the higher-level characteristic will be within its tolerance is wrong. Going back to the example of the quartz watch, if the accuracy tolerance is ± 3.5 seconds per month, that does not mean that the tolerances of the quartz oscillator and the other components are designed to bring the value within 3.5 seconds. After the watch is assembled, its accuracy is measured, and if it is outside the adjustment limit (± 3.5 seconds) it is adjusted (calibrated, reworked) by changing a capacitor parameter. If the loss A from rejecting a quartz oscillator is ¥ 800 and the cost A_0 of adjusting the capacitor is ¥ 600, then with 3.5 seconds as the adjustment limit Δ_0, the tolerance Δ for the accuracy of the quartz oscillator comes out to be ± 4.0 seconds, from the following formula.

$$\Delta = \sqrt{\frac{A}{A_0}} \times \Delta_0 = \sqrt{\frac{800}{600}} \times 3.5 = 4.0 \text{ (sec)}$$

......(D-3.1)

Even when the quartz oscillator is within its specification, the watch may be outside the adjustment limit.

S: If the idea of tolerance allocation is wrong, that explains why the tolerances of the components on drawings from American firms sometimes add up to more than the tolerance of the objective characteristic. I had always been puzzled at that, but now I see that there is nothing wrong with it.

G: That's right. The tolerances should be determined considering each component characteristic value in relation to the higher-level characteristic value separately. In the next chapter we will begin to see how to vary a group of system parameters simultaneously and study their effect on a higher-level characteristic, but even in that case, the tolerances of the parameters should be set individually using formula 3.4.

S: Even when there is a simple and inexpensive way to adjust the value of the higher-level characteristic?

G: If it cannot be adjusted, the product has to be scrapped. In an assembled product, that will cost more than any of the individual system components, so even if the tolerances are set individually, they should ensure that the higher- level characteristic will satisfy its tolerance. Since the method of deciding the tolerances of the individual system components given above results in a reasonable tolerance for the higher-level characteristic, there is no objection to determining the tolerances individually.

S: Not having to allocate tolerances among system components to make a higher-level characteristic meet its tolerance greatly simplifies the tolerance-setting process. The tolerances can be determined just by finding A and β for each system component.

G: The best way to find β is to study it by varying a number of parameters at once. This will be explained when we come to orthogonal arrays. In general, however, it is enough to set tolerances for the individual parameters separately.

4 | TOLERANCE DESIGN AND EXPERIMENTAL DESIGN

Linear and Quadratic Coefficients

A method of determining the tolerances of the initial value and rate of deterioration of a characteristic value of a component that serves as a contributing parameter (system element) was given in the last chapter. The effect of a unit change in the lower-level characteristic value on the higher- level characteristic value was expressed by the linear coefficient β. In general, the coefficients of the linear and quadratic terms are found as derivatives. A more rational method of finding them by techniques of experimental design is described below.

Expansion by Orthogonal Polynomials

An orthogonal expansion provides an effective means of evaluating the influence a contributing variable has on the objective characteristic. Orthogonal expansion by Chebyshev polynomials is particularly useful in tolerance design.

In the past, designers frequently used Taylor expansions to calculate the functional effect of variation in a factor. For example, the exponential function with variable A has the well-known Taylor expansion:

$$e^A = 1 + \frac{A}{1!} + \frac{A^2}{2!} + \cdots + \frac{A^5}{5!} + \cdots \qquad \ldots\ldots(4.1)$$

Table 4.1 shows the error at steps of 0.5 as A varies over the interval $(-2, 2)$ when the function is approximated by its Taylor expansion out to the first through fifth terms. The column headed "3rd," for example, gives the differences between the true value of the function and the value of the following approximation, calculated at intervals of 0.5, starting at -2:

$$F_3(A) = 1 + A + \frac{A^2}{2!} + \frac{A^3}{3!} \qquad \qquad ......(4.2)$$

The error variation in Table 4.1 is the sum of the squares of the errors, and there are nine degrees of freedom (the number of points in the interval). The error variation if the expansion is taken out to the third-degree term is 1.450432, and the error variance is the error variation divided by the number of degrees of freedom (9), or 0.161159.

Table 4.1 Error in Taylor Expansion

A	e^A	Size of error				
		1st	2nd	3rd	4th	5th
-2.0	0.135	-1.135	0.865	-0.468	0.190	-0.068
-1.5	0.223	-0.723	0.402	-0.161	0.050	-0.013
-1.0	0.368	-0.368	0.132	-0.036	0.008	-0.001
-0.5	0.607	-0.107	0.018	-0.003	0.000	0.000
0.0	1.000	0.000	0.000	0.000	0.000	0.000
0.5	1.649	-0.149	-0.024	-0.003	-0.001	0.000
1.0	2.718	-0.718	-0.218	-0.051	-0.010	-0.001
1.5	4.482	-1.982	-0.857	-0.294	-0.084	-0.020
2.0	7.389	-4.389	-2.389	-1.056	-0.389	-0.122
Error variation		25.687197	7.417447	1.450432	0.197142	0.020079

For an orthogonal expansion, the size of the error variation can be found by constructing an analysis-of-variance table as follows. Treating the values 0.135, 0.223, ..., 7.389 of the exponential function at the points where the factor (variable) A equals -2.0, -1.5, ..., 2.0 as data for nine levels of A, we use the orthogonal polynomials for $k = 9$. Writing b_i for the coefficient of the i-th term, w_i from reference (1), the variation for a fifth-degree expansion is:

$$S_{b0} = \frac{(A_1 + A_2 + \cdots + A_9)^2}{9}$$

$$= \frac{(0.135+0.223+\cdots+7.389)^2}{9} = \frac{18.571^2}{9}$$

$$= 38.320226 \quad (f=1) \qquad\qquad \dots\dots(4.3)$$

$$S_{b1} = \frac{(W_1A_1+W_2+A_2+\cdots+W_9A_9)^2}{r\ (\lambda^2 S)}$$

$$= \frac{(-4\times0.135-3\times0.223-\cdots+4\times7.389)^2}{1\times60} = \frac{47.534^2}{60}$$

$$= 37.659604 \quad (f=1) \qquad\qquad \dots\dots(4.4)$$

$$S_{b2} = \frac{(28\times0.135+9\times0.223+\cdots+28\times7.389)^2}{1\times2772}$$

$$= 9.300780 \quad (f=1) \qquad\qquad \dots\dots(4.5)$$

$$S_{b3} = \frac{(-14\times0.135+7\times0.223+\cdots+14\times7.389)^2}{1\times990}$$

$$= 1.022418 \quad (f=1) \qquad\qquad \dots\dots(4.6)$$

$$S_{b4} = \frac{(14\times0.135-21\times0.223-\cdots+14\times7.389)^2}{1\times2002}$$

$$= 0.059226 \quad (f=1) \qquad\qquad \dots\dots(4.7)$$

$$S_{b5} = \frac{(-4\times0.135+11\times0.223-\cdots+4\times7.389)^2}{1\times468}$$

$$= 0.001908 \quad (f=1) \qquad\qquad \dots\dots(4.8)$$

Let S_T be the sum of all the squares:

$$S_T = 0.135^2 - 0.223^2 + \cdots + 7.389^2$$
$$= 86.34197 \quad (f=9) \qquad\qquad \dots\dots(4.9)$$

$$S_e = S_T - (S_{b0} + S_{b1} + \cdots + S_{b5})$$
$$= 0.000034 \quad (f=3) \qquad\qquad \dots\dots(4.10)$$

This gives the analysis of variance shown in Table 4.2.
Tables 4.1 and 4.2 can be used to compare the size of the error in the Taylor expansion and the expansion by orthogonal polynomials. For the third-degree expansion, for example, the Taylor series gives an

error variance of 0.161159, as noted above, while for the orthogonal expansion, the figure is:

$$S_e = S_{b4} + S_{b5} + S_e = 0.061168 \ (f = 5) \qquad ......(4.11)$$

$$V_e = S_e / 5 = 0.012234 \qquad ......(4.12)$$

This error variance is only 1/13.2 as large as the other.

$$\frac{0.161159}{0.012234} \fallingdotseq 13.2 \qquad ......(4.13)$$

If the total squared error is compared, the approximation by orthogonal polynomials is 23.7 times closer. We will therefore examine quality design methods that make use of orthogonal polynomials to approximate arbitrary functions.

In tolerance design, there are usually two or three levels. Table 4.3 gives orthogonal polynomials for up to five levels. There would almost never be as many as nine levels. See chapter 21 in reference (1).

Table 4.2 Analysis of Variance

Factor	f	S	ρ (%)
b_0	1	38.320226	44.370488
b_1	1	37.659604	43.905561
b_2	1	9.300780	10.769242
b_3	1	1.022418	1.183832
b_4	1	0.059226	0.068564
b_5	1	0.001908	0.002197
e	3	0.000034	0.000116
T	9	86.364197	100.000000

To estimate the i-th-coefficient b_i when there are k levels, the weighting factors W_1, W_2, and so on in the b_i column are used.

$$\text{Estimated} \ ' \ \hat{b}_i = \frac{W_1 A_1 + \cdots + W_k A_k}{r \cdot \lambda S \cdot h^i} \quad (i = 1, 2, \cdots, 5)$$

$$\text{Variation} \ ' \ S_{bi} = \frac{(W_1 A_1 + \cdots + W_k A_k)^2}{r \cdot \lambda^2 S} \quad (i = 1, 2; \cdots, 5)$$

The variance is $o^2 + rSh^{2i}b_i^2$ (where r is the iteration, and h is the interval).

Table 4.3 Orthogonal Polynomials

Number of levels / Coefficient	k=2 b_1	k=3 b_1	k=3 b_2	k=4 b_1	k=4 b_2	k=4 b_3	k=5 b_1	k=5 b_2	k=5 b_3	k=5 b_4
W_1	-1	-1	1	-3	1	-1	-2	2	-1	1
W_2	1	0	-2	-1	-1	3	-1	-1	2	-4
W_3		1	1	1	-1	-3	0	-2	0	6
W_4				3	1	1	1	-1	-2	-4
W_5							2	2	1	1
$\lambda^2 S$	2	2	6	20	4	20	10	14	10	70
λS	1	2	2	10	4	6	10	14	12	24
S	$\dfrac{1}{2}$	2	$\dfrac{2}{3}$	5	4	$\dfrac{9}{5}$	10	14	$\dfrac{72}{5}$	$\dfrac{288}{35}$
λ	2	1	3	2	1	$\dfrac{10}{3}$	1	1	$\dfrac{5}{6}$	$\dfrac{35}{12}$

Simple Example of a Two-Way Layout

This problem appears in an experimental design course: If alternating current with a voltage of 100 V and frequency of 50 Hz is applied to a circuit with a resistance of R ohms and self-inductance of L henrys, the resulting current flow in ampere units is given by the following formula.

$$y = \frac{100}{\sqrt{R^2 + (2\pi \times 50L)^2}} \qquad \text{.....(4.14)}$$

Under the consumer's conditions of use, the tolerance range of the current value of this circuit is 10.0 ±4.0 amperes. The loss A incurred through after-service and the like when the specification is not met is ¥ 15,000. The monthly production volume is 200,000 units. Due to the use of second-rate components in the circuit at present, the statistics for the resistance R and self- inductance L are:

Resistance R (ohms)   Mean $m_R = 9.92$

Standard deviation $o_R = 0.300$

Self-inductance L (millihenrys)   Mean $M_L = 4.0$
Standard deviation $\sigma_L = 0.80$

The standard deviations include variation caused by 10 years of deterioration.

In many countries this type of problem is solved by two methods:

(1)   Taylor expansion

(2)   Simulation

The Taylor-expansion method is difficult and unattractive -- unattractive because it is hard to evaluate the error caused by omitting the higher-order terms. Simulation is inefficient. A better way is to use experimental design.

Switching from the symbols R and L to A and B, we can write the three levels of the resistance A and self-inductance B of the coil as follows.

encompasses most of details no matter how skewed)

$$A_1 = 9.920 - \sqrt{\frac{3}{2}} \times 0.300 = 9.552$$

$$A_2 = 9.920 \qquad\qquad\qquad\qquad ......(4.15)$$

$$A_3 = 9.920 + \sqrt{\frac{3}{2}} \times 0.300 = 10.288$$

$$B_1 = 0.00400 - \sqrt{\frac{3}{2}} \times 0.00080 = 0.00302$$

$$B_2 = 0.00400 \qquad\qquad\qquad\qquad ......(4.16)$$

$$B_3 = 0.00400 + \sqrt{\frac{3}{2}} \times 0.00080 = 0.00498$$

For B, the units have been converted from millihenrys to henrys. For the nine combinations of the levels of A and B, the current value can be found from formula 4.14. For example, the current y_{11} for the combination A_1B_1 is:

$$y_{11} = \frac{100}{\sqrt{A_1^2 + (2\pi \times 50 \times B_1)^2}} = \frac{100}{\sqrt{9.552^2 + (314.16 \times 0.00302)^2}}$$

$$= 10.42 \qquad\qquad\qquad\qquad ......(4.17)$$

This and the values similarly obtained for the other combinations are listed in Table 4.4. The data are given as deviations from the target of 10 amperes.

Table 4.4 Two-Way Layout Data

	B_1	B_2	B_3	Total
A_1	0.42	0.38	0.33	1.13
A_2	0.03	0.00	−0.04	−0.01
A_3	−0.32	−0.35	−0.39	−1.06
Total	0.13	0.03	−0.10	0.06

(1) Perform an analysis of variance by resolving into orthogonal polynomial components. Find the variations to five decimal places.
(2) Find the loss due to variability of the output current, and calculate whether it would be advantageous to change to first-rate components which have only half the variability and deterioration. The halving of variability and deterioration reduces the first- and second-order contribution ratios to 1/4 and 1/16 respectively, but the price of first-rate components is ¥ 12 higher for the resistor and ¥ 100 higher for the coil.

Analysis of Variance

We would like to know whether variation of the resistance R or inductance L causes more variation in the output current, and how much larger the effect is. Equation 4.14 does not yield this quantitative information immediately, but Table 4.4 shows the output current obtained as in 4.17 for the nine combinations of the three levels of R and L. We can now use orthogonal polynomials to calculate the variation for the following factors (components of the polynomials):

m = general mean; a constant term corresponding to the mean
A_l = linear effect of A
A_q = quadratic effect of A
B_l = linear effect of B
B_q = quadratic effect of B
$A_l \times B_l$ = product of linear effects of A and B
e = so-called error term, covering all the higher-degree terms

The variation S_m of the general mean is the square of the total divided by the number of data.

$$S_m = \frac{0.06^2}{9} = 0.00040$$

......(4.18)

S_{Al} is found as follows using the weights 1, 0, and -1 for the first term b_1 for the case of three levels ($k = 3$) in Table 4.3:

$$S_{Al} = \frac{(-1 \times A_1 + 0 \times A_2 + 1 \times A_3)^2}{r \lambda^2 S} \qquad ......(4.19)$$

Here A_1, A_2, and A_3 are the totals for the three levels of A. From Table 4.4 their values are 1.13, -0.01, and -1.06. The symbol r, the so-called iteration, indicates the number of calculated values (data) figuring in the sums A_1, A_2, and A_3. Here $r = 3$. The value of $\lambda^2 S$ is the value given in Table 4.3 for $k = 3$ and b_1. It is also the sum of the squares of the weights -1, 0, and 1.

$$\lambda^2 S = (-1)^2 + 0^2 + 1^2 = 2$$

Inserting these values into equation 4.19, we have:

$$S_{Al} = \frac{[-1 \times 1.13 + 1(-1.06)]^2}{3 \times 2} = 0.79935 \qquad ......(4.20)$$

S_{Aq} is found using the weights 1, -2, and 1 given in Table 4.3 for $k = 3$ and b_2, with $r = 3$ and $\lambda^2 S = 6$.

$$S_{Aq} = \frac{[1 \times 1.13 - 2(-0.01) + 1 \times (-1.06)]^2}{3 \times 6}$$

$$= 0.00045 \qquad ......(4.21)$$

Similarly,

$$S_{Bl} = \frac{[-1 \times 0.13 + 1 \times (-0.10)]^2}{3 \times 2}$$

$$= 0.00882 \qquad ......(4.22)$$

$$S_{Bq} = \frac{[1 \times 0.13 - 2 \times 0.03 + 1 \times (-0.10)]^2}{3 \times 6}$$

$$= 0.00005 \qquad ......(4.23)$$

The variation $S_{Al \times Bl}$ of the product of the linear effects of A and B (the interaction $A_l \times B_l$) is found as follows. Using the weights -1, 0, and 1 for the linear effect of B on each level of A, the following contrasts $L(A_1)$, $L(A_2)$, and $L(A_3)$ can be found.

$$L(A_1) = -1 \times 0.42 + 0 \times 0.03 + 1(-0.32)$$
$$= -0.74 \qquad \qquad ......(4.24)$$

$$L(A_2) = -1 \times 0.38 + 0 \times 0.00 + 1(-0.35)$$
$$= -0.73 \qquad \qquad ......(4.25)$$

$$L(A_3) = -1 \times 0.33 + 0 \times (-0.04) + 1(-0.39)$$
$$= -0.72 \qquad \qquad ......(4.26)$$

Multiplying $L(A_1)$, $L(A_2)$, and $L(A_3)$ by the linear term weights $-1, 0$, and 1 for A, we get:

$$S_{Al \times Bl} = \frac{[-1 \times (-0.74) + 0(-0.73) + 1(-0.72)]^2}{2 \times 2}$$
$$= 0.00010 \qquad \qquad ......(4.27)$$

The sum of the squares S_T is:

$$S_T = 0.42^2 + 0.38^2 + \cdots + (-0.39)^2$$
$$= 0.8092 \qquad \qquad ......(4.28)$$

The variation of the errors S_e is:

$$S_e = S_T - (S_m + S_{Al} + \cdots + S_{Al \times Bl})$$
$$= 0.8092 - (0.00040 + 0.79935 + \cdots + 0.00010)$$
$$= 0.00003 \qquad \qquad ......(4.29)$$

From these figures we can perform analysis of variance as in Table 4.5. For details, see chapter 21 in reference (1).

Table 4.5 Analysis of Variance

Source		f	S	V	ρ (%)
m		1	0.00040	0.00040	0.045
A	l	1	0.79935	0.79935	98.778
	q	1	0.00045	0.00045	0.051
B	l	1	0.00882	0.00882	1.086
	q	1	0.00005	0.00005	——
$A_l \times B_l$		1	0.00010	0.00010	——
e		3	0.00003	0.00001	——
(e)		(5)	(0.00018)	(0.000036)	0.040
T		9	0.80920	0.08991	100.000

In this type of analysis of variance, the designer has complete freedom in deciding how to deal with the error variance, just as there is complete freedom in deciding how far to take the expansion of the exponential function before treating the rest of the terms as error. The problem is that the magnitude of the error must be evaluated correctly and precisely when the decision is made. The model is unrestricted, but the error must be evaluated precisely.

Comparing the size of the variances, let us decide to leave m, A_l, A_q, and B_l and pool the rest of the terms with the error variation. The pooled error variation is (e), there are five degrees of freedom, and the variation is 0.00018. The error variance is:

$$V_e = \frac{0.00018}{5} = 0.000036 \qquad \text{......(4.30)}$$

This was used in calculating the percent contribution. For example:

$$\rho_m = \frac{S_m - V_e}{S_T} \times 100 = \frac{0.00040 - 0.000036}{0.80920} \times 100 = 0.045 \quad (\%) \qquad \text{......(4.31)}$$

When second-rate components are used, the variance will be:

$$V_T = \frac{S_T}{9} = \frac{0.80920}{9} = 0.08991 \qquad \text{......(4.32)}$$

The percent contribution is the magnitude of the effect of m, A_l, A_q, B_l, or the remaining terms (higher-order terms, interactions, model error, etc.) in relation to the size of this variance.

Tolerance Design

In tolerance design, decisions are made as to how much variability to allow in component parts. The necessary information is the loss function and the amount of variability of each grade of component. (The variability is given by the standard deviation which includes deterioration and the change affected by external factors.)

From the problem, the loss function L for the output current is:

$$L = \frac{15000}{4.0^2}\sigma^2$$

$$= ¥ 937.5 \times \sigma^2 \qquad \text{......(4.33)}$$

The loss using second-rate components is therefore:

$$L = 937.5 \times V_T$$
$$= 937.5 \times 0.08991$$
$$= ¥\,84.3 \qquad\qquad\qquad ......(4.34)$$

There is an annual loss of ¥ 16,860,000 due to variability. If first-rate components are used, the variability and deterioration will be reduced by half, which means that their standard deviation will be reduced by half, which means in turn that the interval h in equations 4.15 and 4.16 will be reduced by half.

In Table 4.5, the expected value of the variance with regard to the linear and quadratic terms is:

$$\sigma^2 + r \cdot S \cdot h^4 \, b_1^2$$
$$\sigma^2 + r \cdot S \cdot h^4 \, b_2^2 \qquad\qquad ......(4.35)$$

The net variation when first-grade components are used will therefore be reduced to 1/4 in the linear terms and 1/16 in the quadratic terms.

The calculations for the resistance R are shown below.

(a)For second-rate components:

$$L = 937.5 \times V_T \times (\rho_{Al} + \rho_{Aq})/100$$
$$= 937.5 \times 0.08991 \times (0.98778 + 0.00051)$$
$$= ¥\,83.30 \qquad\qquad ......(4.36)$$

(b)For first-rate components:

$$L = 937.5 \times 0.08991 \times \left(0.98778 \times \frac{1}{4} + 0.00051 \times \frac{1}{16}\right)$$
$$= ¥\,20.82 \qquad\qquad ......(4.37)$$

From the above, the quality improvement is:

$$83.30 - 20.82 = ¥\,62.48 \qquad\qquad ......(4.38)$$

Since the cost of the improvement is ¥ 12, the net gain is ¥ 50.48, or roughly ¥ 10 million annually.

Table 4.6 summarizes the results of these calculations and similar calculations for the inductance. It indicates that first-rate resistors and second-rate coils should be used. Although this increases the cost by ¥ 12, the first-rate resistor yields improved quality and the contribution level of the general mean ρ_m becomes 0, so under optimum conditions there is a gain of:

$$-12 + 62.48 + 48.3 \times 0.00045 = ¥\,50.52$$

Table 4.6 Tolerance Design

Component	Grade	Cost	Quality	Total	
Resistance	2	Base cost	¥ 83.30	¥ 83.30	
	1	+ ¥ 12	¥ 20.82	¥ 32.82	Loss-
Inductance	2	Base cost	¥ 0.92	¥ 0.92	add
	1	+ ¥ 100	¥ 0.23	¥ 100.23	For Total Los

The annual gain is ¥ 10,000,000.

In this chapter two-way arrays were used for tolerance design, but usually there are more system parameters, and orthogonal arrays are used.

PROBLEM

When an object with mass $m = 0.2\ kg$ is projected by a force of F newtons at an angle of elevation a in a direction of θ, the distance y (in meters) that it travels over the plane is given by the formula:

$$y = \frac{1}{g} \times \left(\frac{F}{m}\right)^2 \sin 2\alpha$$

$$= \frac{1}{9.807} \times \left(\frac{F}{0.2}\right)^2 \sin 2\alpha$$

$$= 2.549 F^2 \sin 2a$$

The mean values of the force F and angle a are $80\ N$ and $15°$. Their standard deviation σ_F and σ_a are:

$$\sigma_F = 5.0\ N$$
$$\sigma_a = 2.0°$$

Three-level factors A and B for the variables F and a are set as follows.

$$A_1 = 80.0 - \sqrt{\frac{3}{2}} \times 5.0 = 73.9\ N$$

$$A_2 = 80.0\ N$$

$$A_3 = 80.0 + \sqrt{\frac{3}{2}} \times 5.0 = 86.1\ N$$

$$B_1 = 15° - \sqrt{\frac{3}{2}} \times 2.0° = 12.55°$$

$$B_2 = 15°$$

$$B_3 = 15° + \sqrt{\frac{3}{2}} \times 2.0° = 17.45°$$

(1) Find the distances of travel for a two-way layout of A and B.
(2) Perform analysis of variance.
(3) If reducing the variability of the force F by half would cost an extra ¥ 12 million, and reducing the variability of the angle a by half would cost an extra ¥ 30 million, find the optimum solution. When the error in the distance of travel y is 50 m, the loss is ¥ 8,000 per shot and total shot number is 50.

DISCUSSION

Orthogonal Polynomials

S (Student): How were the orthogonal polynomials derived, and what is their general form?

G (Genichi): There are many types of orthogonal polynomials, but the only ones that can be used for approximating function values or experimental values at k equally-spaced levels are the Chebyshev polynomials given in Table 4.3. First the k values are approximated by a constant, then their differences with that are approximated by the linear polynomial, and so on, so the lower-degree polynomials have priority.

S: In that case, if a characteristic value y is, say, a quadratic function of A,

$$y = bA^2$$

then if orthogonal polynomials are used to approximate it, the constant and linear effects will also be significant.

G: Significance is nearly meaningless in statistical testing, but the constant and linear polynomials will have a large effect, so their percent contribution will be high.

S: If the orthogonal polynomial expansion can be used even in that case, apparently it does not matter if the model is incorrect.

G: True. Similarly, the exponential function is not a polynomial, but it can be approximated by a polynomial.

S: What is the general form of the orthogonal polynomials?

G: In practice, we never go past the cubic polynomial; the higher-degree polynomials are of only theoretical interest. If there are k levels spaced at intervals of h, the formula through the cubic term is as follows:

$$b_0 + b_1 (A - \overline{A}) + b_2 \left[(A - \overline{A})^2 - \frac{k^2 - 1}{12} h^2 \right]$$

$$+ b_3 \left[(A - \overline{A})^3 - \frac{3k^2 - 7}{20} (A - \overline{A}) h^2 \right] + \cdots$$

The size of the variation of each term is given by formula D-4.1:

$$r \cdot S \cdot h^{2i} b_i^2 \qquad\qquad\qquad \text{......(D-4.1)}$$

S: So if the variability is reduced by half, the size of the i-th term is reduced by $(1/2)^{2i}$.

G: That's right. Even when the higher-degree terms are large, they quickly approach zero if the size of the variability is reduced. Accordingly, when the second-degree term is not extremely large, the higher-degree terms can often be omitted from the percent contribution.

Finding Tolerances

S: Earlier we saw that each of the system elements can be calculated separately in tolerance design.

G: True. Allocating tolerances by dividing the tolerance of the objective characteristic among the tolerances of the system parameters is incorrect. In the case of a circuit, the tolerance for each of the circuit elements should be determined separately. In analysis of variance to find the size of the effect of each of those elements, however, it is best to treat all the elements at once in the calculation or experiment. This will be explained in the next chapter.

S: If the tolerances can be determined individually, why do the different elements have to be studied together?

G: Because the impact a change in an element has on the higher-level characteristic depends on the values of the other elements at that time.

S: In other words there is a mutual interaction. I've been told that when there is interaction, it should be calculated. Is that true?

G: Finding the interaction can be extremely difficult. The reason for using orthogonal arrays and pooling all the error factors is to evaluate the interactions together with the higher-degree terms. If the size of the higher-degree terms, including the interactions, can be evaluated, then an approximation can be used within that range. In effect, the purpose is to find the size of the error when the function is expressed in terms of the main effects. But we shall look at the role of orthogonal arrays in Chapter 6.

5

OFF-LINE AND ON-LINE QUALITY CONTROL

Variability Due to Error Factors, and Countermeasures

The factors that cause variability in product functions are called error factors or noise. Such factors are the reason, for example, that the brightness of flourescent lamp varies with the power supply voltage and deteriorates over time. There are three main types of noise:

(1) External noise -- variables in the environment or conditions of use that disturb the functions of a product. Temperature, humidity, dust, and individual human differences are examples of external noise.

(2) Deterioration noise or internal noise -- changes that occur when a product deteriorates during storage or wears out during use, so that it can no longer achieve its target functions.

(3) Variational noise or unit-to-unit noise -- differences between individual products that are manufactured to the same specifications.

To overcome the problems of variability, design and production engineering departments use methods of off-line quality control, while production departments use methods of on-line quality control. Both methods are discussed in this chapter.

Off - Line Quality Control (Quality Engineering)

A firm decides on a set of target functions, prepares the necessary specifications and drawings, and begins to manufacture the product.

Some of the manufactured units may possess the target functions while others do not; this is due to unit-to-unit noise. The product may break down after prolonged use; its functions have been impaired by deterioration affected by noise. The product may function well under normal conditions but not under high temperature or humidity, or when the power supply voltage is 20% off nominal; these are problems of external noise.

Good functional quality means little functional variation from any of the above types of noise -- a product which functions as intended under a wide range of conditions for the duration of its design life. The ideal in functional quality is for the functions to remain normal despite fluctuations in temperature, humidity, supply voltage, and other external (environmental) factors, even when components and materials degrade or wear down during long use, and notwithstanding unit-to-unit variability. Quality with respect to an objective function can be measured as the degree of variation from the target functional value, the nominal or ideal value determined in the specifications.

In the late 1940s, the author assisted a firm in experiments with caramel candy. At the time, the plasticity (chewability) of the firm's caramel varied greatly with the ambient temperature, as shown by curve A_1 in Figure 5.1. The firm hoped to reduce this temperature dependence. Caramel is manufactured from a mixture of more than ten separate ingredients, so there were many factors to experiment with in search of a temperature-flat plasticity curve. Although the result fell short of complete success, a curve roughly like A_2 was achieved. The firm now had a more "reliable" caramel that is more resistant to external noise.

Internal noise is a variation in the internal constants of the product itself, causing variation in its objective functions. The variation may be over time (degradation) or space (unit to unit). If a product maintains its initial performance indefinitely, it has a high degree of stability. If all units give exactly the same performance, the manufacturer earns a reputation for reliability and uniformity.

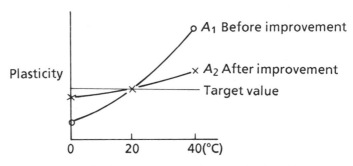

Figure 5.1   Ambient Temperature and Plasticity

The function of a power circuit in a color television set is to convert 100-V AC input into 115-V DC output. If the power circuits in all sets manufactured maintained 115-V output under all conditions, their voltage quality would be perfect. The following noise factors, however, may cause the output to deviate from its 115-V target.

(1) External noise: all variations in environmental conditions such as temperature, humidity, dust, and input voltage.

(2) Internal noise: changes in the components and materials of the circuit. For example, after 10 years the resistance of a resistor may have increased by 10%, or h_{FE} and other transistor parameters may vary with time.

(3) Unit-to-unit noise: differences between the individual manufactured units, causing different output voltages from the same input voltage.

Assuring functional quality means finding means to reduce the effect of these three types of noise. The most important means is design, which is an aspect of off-line quality control. Design measures can be taken in three steps:

(1) System design (or primary design): functional design, focused on the pertinent technology.

(2) Parameter design (or secondary design): a means of both reducing cost and improving quality, making effective use of experimental design methods.

(3) Tolerance design (or tertiary design): a means of controlling causes at increased cost. Experimental design methods can be used.

(1) System Design (Primary Design, Functional Design)

This is the step in which one surveys the pertinent technology and asks, for example, what kinds of circuits could be used to convert alternating current to direct current, or what reaction processes could be used to produce a desired chemical product. It is a search for the best available technology.

Automatic control systems can be considered in this step. For the TV power circuit, one might think of a system that measures the instantaneous output voltage and corrects deviations from the 115-V target automatically by controlling a circuit parameter, perhaps by changing the value of a variable resistance element. It is difficult, however, to control deterioration and other changes in the automatic control system itself, and its inclusion adds cost. This is not the way to design an inexpensive, stable circuit.

Although system design is extremely important, it is not possible to study all possible systems, so a small number, say one to three, must be selected on the basis of assessment or gut feeling. Such selections are repeated through a large number of product development steps.

For a television set these would be:

(a) Total system design
(b) Subsystem design
(c) Unit and component design
(d) Element development
(e) Material development

Each of the above steps has its own system design, parameter design, and tolerance design. It is hard to make accurate judgments in system development because so much depends on the system designer's predictions. Likewise, system design is a technology issue. It is in parameter and tolerance design that experimental design methods have a role to play.

(2) Parameter Design (Secondary Design)

After the system design is settled, the optimum levels of the individual system parameters have to be decided. There has been a tendency to neglect parameter design research, particularly in developing countries. Research and development engineers in those countries take the pertinent technology from the literature and other sources, design what seems to be the optimum power circuit, build a prototype, and apply 100-V alternating current to it. If the resulting output voltage is the target 115 V, the design is deemed a success. If it deviates from 115 V, they adjust it by altering one parameter value in the circuit. They are like a dyer given a color sample to match. Using all his professional knowledge, he makes a test dye; if the result matches the sample, the dye is a success. If the hue or saturation deviates from the target value, he changes the mixing ratios or dyeing conditions until the target is matched. If there are three target characteristics, the dyer can match them by experimenting with just three independent control variables, one with a large impact on each characteristic.

This type of adjustment to a target value (or to a set of target values -- the principle is the same) belongs in the manufacturing process or the field, not in the design process. The term for it is modification, or calibration.

Engineers in developing countries rarely do any real design work. They simply make adjustments, working at the level of the dyer. For example, suppose that the system is a power circuit, A is the h_{FE} parameter of a transistor element, B is the resistance of a resistor element, and these factors affect the output voltage as shown in Figure 5.2. A prototype circuit is built with $A_1 = 10$, and 100-V alternating current is applied to it; the resulting output is only 90 V. From Figure 5.2, the parameter with the greatest effect on the output voltages appears to be the transistor's h_{FE}, so the designer tries increasing it to 30 and 50, getting the output voltages shown on the left in Figure 5.2.

These values could be obtained from theory, or by experiment if there is no applicable theory. The output voltage curve meets the target value of 115 V at the point $A' = 20$. To use this for the value of h_{FE}, however, would be a bad mistake.

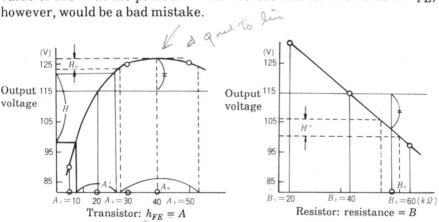

Figure 5.2 Parameter Values and Output Voltage

The reason is that the goal of the design is to eliminate variability while using inexpensive components. There may be other target characteristics such as current value, but here we will consider only the output voltage. Transistors and resistors with reliable parameter values that do not deteriorate are expensive. The least expensive element costs less than one-tenth of what the highest-grade element costs. Suppose that third-grade elements are being considered as the inexpensive choice. If the design life is 10 years, the transistor's h_{FE} will vary and deteriorate from its nominal value by about $\pm 30\%$. If the parameter value $A' = 20$ is selected, and the inexpensive transistor with a variability of $\pm 30\%$ is used, its h_{FE} will vary between 14 and 26, so the output voltage will vary between 98 V and 121 V, a range of 23 V.

If instead of A' the value $A_0 = 40$ were selected and again the parameter h_{FE} varied by $\pm 30\%$, then even though h_{FE} would range between 28 and 52, the range of variation of the output voltage would be only 5 V, from 122 V to 127 V. The objective characteristic (the output voltage) would be quite steady despite the use of inexpensive components. Unfortunately, it would also be 10 V over target, but that could be corrected by changing the resistance parameter, which has a linear effect on the output voltage, as shown on the right in Figure 5.2. Only one variable is needed to correct deviation from a target value. The author uses the term adjustment or signal factor to describe a variable used for this purpose.

When a parameter has a nonlinear influence on the objective characteristic, like the parameter h_{FE}, the effect of its own variability can be reduced by choosing the value at the peak of the curve, A_0 in Figure 5.2. This is an advanced design technique; it both reduces cost and improves quality. Engineers have long been taught to use nonlinear relationships. Still, this does not automatically mean that $A = A_0$ is the best choice. Although it reduces the effect of variation in h_{FE}, it might amplify the effect of variation in other parameters. Design research should take changes in environmental conditions and the other parameters into account. It is here that orthogonal arrays and accumulation methods must be used. For further details, see reference (1).

When the goal is to design a product (or process) with high stability and reliability, parameter design is the most important step. This is the step in which nonlinearity is used, as in the factor A. This is the step in which we find the combination of parameter levels that reduces the effect not just of internal noise but of all noises while maintaining a constant output voltage. It is the central step in design research, the answer to the call to design a product or process that exhibits high reliability under a wide range of conditions, despite the use of inexpensive, highly variable, and easy to deteriorate materials and parts.

These are exactly the aims of experimental design. Studying a large number of factors and selecting the optimum combination of factor levels is basically an application of nonlinearity, even when not recognized as such. The key to success in experimental design is to select objective characteristics and cite factors with the above aims in mind. In the developing countries, where raw materials and components tend to be highly variable, this is the most important quality control measure; parameter design must attain a much higher degree of stability than in the industrialized countries. If the product can be designed so that its output characteristics are resistant to both external and internal noise, then it will function satisfactorily despite variability in its component parts, and its cost will be low. This is the ideal way to deal with all three sources of noises.

The major usefulness of experimental design is in the secondary design process of finding the optimum combination of parameter levels. The reason that some manufacturers in developing countries turn out products of inferior quality despite their thorough knowledge of the literature and use of industrialized countries' technology is that their engineers do not experiment. They blindly accept the parameters given in the literature or their advanced partners' specifications.

(3) Tolerance Design (Tertiary Design)

After the system has been designed and the nominal mid-values of its parameters determined, the next step is to set the tolerances of the parameters. Environmental factors must be considered together with the system parameters. We assign all these factors to an orthogonal array with levels that reflect their variability from their mid-values to find the extent of their impact on the output characteristics. Narrow tolerances can be given to the noise factors with the greatest influence. The methodology is different than in parameter design, however. Now we are attempting to control the error factors, and keeping them within narrow tolerances will drive the cost up. This is why every possible effort should be made to incorporate quality design measures in the parameter design step. Narrow tolerances should be the weapon of last resort, to be used only when parameter design gives insufficient results, and never without careful evaluation of the loss due to variability. Cost calculations determine the tolerances.

Tolerance design was covered in Chapters 2 to 4. For the use of orthogonal arrays, see Chapter 6.

Design of Production Processes

The results of system, parameter, and tolerance design by the design department are passed to the production department in the form of specifications. The production department then designs a manufacturing process that will adequately satisfy these specifications. Process design is also done in three steps:

(1) System design: in which the manufacturing process is selected from knowledge of the pertinent technology, which may include automatic control.

(2) Parameter design: in which the optimum working conditions for each of the component processes are decided, including the optimum materials and parts to purchase. The purpose of this step is to improve process capability by reducing the influence of harmful factors.

(3) Tolerance design: in which the tolerances of the process conditions and sources of variability are set. This is a means of suppressing quality variation by directly removing its cause.

The efficiency of steps (2) and (3) can frequently be raised by means of experimental design. Step (2) is often more important than step (3). In 1953, Ina Seito found that a 5% admixture of a certain type of lime to the raw materials used in manufacturing tile would reduce variability of the fired dimensions by a factor of about 10. The variability was caused by uneven temperature distribution in the tunnel kiln, but Ina's solution did not attack the cause; it simply diminished its influence. This was an application of nonlinearity, which is the basic aim of experimental design. It was highly praised in the book Sangyo Furontea Monogatari (Frontier stories in industry, published by

Diamond), which noted that Ina Seito was the first manufacturer anywhere to develop a calcereous tile. This was the result of parameter design by the design department working on the production process. Contrast this with tolerance design, in which the causes of variability are identified and held within narrow tolerances, usually at high cost.

Production process designs, including the setting of production conditions, are studied by the production engineering department in the test run stage. Hence they are a form of off-line quality control, and should use the methods of experimental design. A simple example of these methods will be given in following paragraph. For further details, see reference (1).

The Tile Experiment

Table 5.1 is the $L_8(2^7)$ orthogonal array, also simply called L_8

Table 5.1 L_8 Orthogonal Array

No.\Col	1	2	3	4	5	6	7
1	1	1	1	1	1	1	1
2	1	1	1	2	2	2	2
3	1	2	2	1	1	2	2
4	1	2	2	2	2	1	1
5	2	1	2	1	2	1	2
6	2	1	2	2	1	2	1
7	2	2	1	1	2	2	1
8	2	2	1	2	1	1	2

The numbers 1 to 8 are called experiment numbers. Each vertical column of the array contains four 1s and four 2s. For a given experiment, in any two columns there are four possible combinations: (1, 1), (1, 2), (2, 1), and (2, 2). If each of these four combinations appears the same number of times in a pair of columns, the columns are said to be balanced, or orthogonal. Select any two of the seven columns in the L_8 array and count the (1, 1), (1, 2), (2, 1), and (2, 2) combinations; you will find that the columns are orthogonal.

Suppose we are studying seven factors (A, B, C, D, E, F, and G), each of which has two levels. (The following technique can also be used when there are fewer than seven factors. See reference (1) if there are more than seven factors, or if some of them have more than two levels.) The seven factors can be assigned to columns 1 through 7 of the L_8 array, and eight experiments can be performed in an order decided by drawing lots. This is what Ina Seito did in its tile manufacturing experiment in 1953.

The tile production process is diagrammed below:

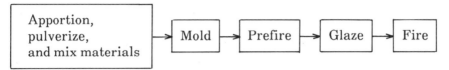

Suggestions for improving the mixture of materials had been tested in small-scale laboratory experiments. The plan was to see if these would work on a 1-ton production scale, and if so, to use them in actual production.

All of the seven factors were related to apportionment of materials. Of the two levels, one was the level in use at the time, and the other was one the experimenters thought might be superior in terms of quality or cost.

Table 5.2   Factors and Levels

Factor	Level 1	Level 2
A: Lime additive content	A_1 = 5%	A_2 = 1% (current)
B: Granularity of additive	B_1 = coarse (current)	B_2 = fine
C: Agalmatolite content	C_1 = 43%	C_2 = 53% (current)
D: Type of agalmatolite	D_1 = current mixture	D_2 = less expensive mixture
E: Charge quantity	E_1 = 1,300 kg	E_2 = 1,200 kg
F: Waste return content	F_1 = 0%	F_2 = 4% (current)
G: Feldspar content	G_1 = 0%	G_2 = 5% (current)

In factor E, productivity rather than quality was the major consideration. If the current 1,200-kg charge could be increased to 1,300 kg without hurting quality, productivity could be increased by nearly 10%. In an experiment intended to raise productivity or reduce costs, one seeks factors that will have a maximum cost or productivity effect but a minimum quality effect. When one hears of a production experiment that saved the manufacturer thousands or even millions of yen, it usually means that the manufacturer succeeded in finding a factor with a large cost impact but no significant quality impact. (The credit in such success stories should go to the experiment, not to experimental design. The true place to use experimental design is not at the factory while production is in progress, but in the product development stage.)

The seven factors in Table 5.2 were assigned to the L_8 orthogonal array as shown on the left side of Table 5.3. The experiments were carried out by preparing eight charges as described in the table, molding them into tiles, and firing them in a tunnel kiln.

Table 5.3 Layout on L_8 Orthogonal Array

Factor	$L_8(2^7)$ array A B C D E F G	Lime additive content (%)	Granu-larity	Agalmatolite content (%)	Type of agalmatolite	Charge quantity (kg)	Waste return content (%)	Feldspar content (%)	Number defective Per 100 tiles on inside
No\Col	1 2 3 4 5 6 7	1	2	3	4	5	6	7	
1	1 1 1 1 1 1 1	5	Coarse	43	Present	1300	0	0	16
2	1 1 1 2 2 2 2	5	Coarse	43	New	1200	4	5	17
3	1 2 2 1 1 2 2	5	Fine	53	Present	1300	4	5	12
4	1 2 2 2 2 1 1	5	Fine	53	New	1200	0	0	6
5	2 1 2 1 2 1 2	1	Coarse	53	Present	1200	0	5	6
6	2 1 2 2 1 2 1	1	Coarse	53	New	1300	4	0	68
7	2 2 1 1 2 2 1	1	Fine	43	Present	1200	4	0	42
8	2 2 1 2 1 1 2	1	Fine	43	New	1300	0	5	26

In experiment No. 1, for each of factors A through G, the number 1 appears in the orthogonal array, so in these experiments level 1 was used for all the factors. This could be described as experiment $A_1 B_1 C_1 D_1 E_1 F_1 G_1$. Specifically, in experiment No. 1 the table says to use a 5% admixture (level A_1) of the lime additive in its coarse-grained form (level B_1), with a 43% (level C_1) agalmatolite content of the present type (level D_1) but no waste return or feldspar (levels F_1 and G_1), in a 1, 300 kg charge (level E_1). The L_8 array can be read as a set of instructions for eight particular experiments that will give a fair comparison of A_1 and A_2, B_1 and B_2, and so forth to G_1 and G_2.

A sample of one hundred tiles was taken under each of the eight experimental conditions. The numbers of tiles that failed to qualify as top-grade are listed in the rightmost column of Table 5.3. A_1 and A_2 were compared by comparing the total number of these defectives in the experiments that used level A_1 (Nos. 1, 2, 3, and 4) with the total for the experiments that used level A_2 (Nos. 5, 6, 7, and 8). These totals were:

$$A_1 = 16 + 17 + 12 + 6 = 51 \qquad \text{......(5.1)}$$
$$A_2 = 6 + 68 + 42 + 26 = 142 \qquad \text{......(5.2)}$$

The percent defective was obtained by dividing the above figures by four:

$$\overline{A}_1 = 12.75\% \qquad \text{......(5.3)}$$
$$\overline{A}_2 = 35.50\% \qquad \text{......(5.4)}$$

This indicated that if the amount of additive were increased from the current 1% to 5%, the percent of defectives would fall from 35.50% to 12.75%.

Similarly, B_1 and B_2 can be compared by comparing the average for the experiments that used level B_1 (Nos. 1, 2, 5, and 6) with the average for the experiments that used level B_2 (Nos. 3, 4, 7, and 8), and the same can be done for the other factors. The results are shown in Table 5.4. The optimum set of conditions is $A_1 B_2 C_2 D_1 E_2 F_1 G_2$.

Table 5.4 Summarized Results

Factor and level	Total number defective	Percent defective
A_1	5 1	1 2 . 7 5
A_2	1 4 2	3 5 . 5 0
B_1	1 0 7	2 6 . 7 5
B_2	8 6	2 1 . 5 0
C_1	1 0 1	2 5 . 2 5
C_2	9 2	2 3 . 0 0
D_1	7 6	1 9 . 0 0
D_2	1 1 7	2 9 . 2 5
E_1	1 2 2	3 0 . 5 0
E_2	7 1	1 7 . 7 5
F_1	5 4	1 3 . 5 0
F_2	1 3 9	3 4 . 7 5
G_1	1 3 2	3 3 . 0 0
G_2	6 1	1 5 . 2 5
Total	1 9 3	2 4 . 1 2

The reason for recommending the use of orthogonal arrays in experiments is that they give highly reliable results for down-stream.

On-line Quality Control

After a production process and operating conditions have been determined, the following sources of product variability remain:
- Variability in materials and purchased components
- Process drift, tool wear, machine failure, etc.
- Variability in execution
- Human error

These sources of variability are dealt with by quality control during normal production, that is, by on-line (real-time) quality control. There are three forms of on-line quality control.
(1) Process diagnosis and adjustment: also known as process control. The process is diagnosed at regular intervals. If it is normal, production is continued. If it is not, the cause is found, and

production is restarted after restoring the process to its original state. Alternatively, preventive adjustments can be made when imminent failure is diagnosed.

(2) Prediction and correction: also known as control. A quantitative characteristic to be controlled is measured at regular intervals, and the measured value is used to predict the (mean) characteristic value of the product if production is continued without adjustment. If the predicted value differs from the target value, the level of a corrective signal factor is modified to reduce the difference. This method is also called feedback or feed foward control. It depends heavily on rational system design.

(3) Measurement and action: also called inspection. Each unit manufactured is measured, and if it is out of specification, it is reworked or scrapped. This method of quality control deals only with the product, while methods (1) and (2) deal mainly with the process.

In managing an automatic control system or robot, it is necessary to check the sensors, that is, the measuring system. Let us consider this from of control.

In the measurement and action method (3) described above, the products are measured and sorted into pass and fail groups, and the corrective action is directed toward the product. If the product were an instrument, this would be equivalent to a pass/fail instrument inspection, and not an in-process instrument calibration. The purpose of calibration is to correct parameter deviation of the instrument while it is in use; the corresponding control method is the prediction-correction method.

If the instrument gradually (or suddenly) begins to exhibit severe deviation, it must be repaired or replaced. This is equivalent to the quality control method of diagnosis and adjustment. The general practice is to repair or replace the instrument if its error at the point of diagnosis exceeds the tolerance of the product characteristic.

If there is a procedure (called diagnosis in on-line quality control) for deciding when calibration is not enough and the instrument must be adjusted (a general term for repaired or replaced), the important factor is not the calibration system but the design of the diagnosis and adjustment system.

We are dealing here only with actions taken on a regular basis at the production site. We are not dealing with what practitioners of the control chart methodology call basic measures, which means finding the cause of variability and removing it permanently (tolerance design in off-line quality control). It is the author's belief that the proper way to design an on-line quality control system is to make sure that the out-of-control region in the control charts is never reached. For specific examples, see reference (2). The object of this section is to give an outline of on-line quality control.

(1) Formulas for Diagnosis and Adjustment, with Examples

A firm manufactures cylinder blocks for truck engines in about 28 process steps. Quality control is needed to see that all of these steps are carried out correctly. Details are given in reference (2). Here we shall describe one of the steps, a reaming process. About ten holes are reamed at once, and all of them have to be good. A departure of 10 microns or more anywhere makes the cylinder block defective and defective cylinder blocks have to be scrapped at a loss A of ¥ 8,000. The cost B of determining whether the holes are true is ¥ 400, and at present every 30th block is diagnosed. The symbol n is used for the diagnostic interval; in the present case, $n = 30$. About 18,000 units were produced during the preceding half year, during which time seven failures were diagnosed. The mean interval between failures $\bar{u}$ is accordingly:

$$\bar{u} = \frac{18000}{7} \fallingdotseq 2,570 \text{ unit}$$

......(5.5)

When a crookedly reamed cylinder or other failure occurs, the process is stopped, the reamer is replaced, then one block is reamed and checked to see if it is good. If it is, production is resumed. The tool cost, labor cost, and other costs associated with stopping the process and replacing the reamer are termed the adjustment cost. In this example the adjustment cost C is ¥ 20,000.

There are three elements in an on-line process adjustment system: the process, the diagnostic method, and the adjustment method. They are characterized by the parameters A, B, C, $\bar{u}$, and l, where l is the time lag and interrelated by the diagnostic interval n. The quality control cost L when the diagnostic interval is n can be derived from process adjustment theory, and is given by the following formula. See reference (2).

$$L = \frac{B}{n} + \frac{n+1}{2} \cdot \frac{A}{\bar{u}} + \frac{C}{\bar{u}} + \frac{l A}{\bar{u}}$$

......(5.6)

In this example, $n = 30$, $A = 8,000$, $B = 400$, $C = 20,000$, $\bar{u} = 2,570$, and $l = 1$, so:

$$L = \frac{400}{30} + \frac{30+1}{2} \cdot \frac{8000}{2570} + \frac{20000}{2570} + \frac{1 \times 8000}{2570}$$

$$= 13.3 + 48.2 + 7.8 + 3.1$$

$$= ¥ 72.5$$

......(5.7)

This is the quality control cost per unit. If the annual production volume is 36,000 units, the cost is ¥ 72.5 × 36,000 = ¥ 2.61 million. A pass/fail quality- control system can be rationalized by reducing the quality control cost L given by the above formula. The cost can be attacked via the specific process technology or via general quality control techniques. Reducing the failure rate, finding simpler diagnostic procedures, and reducing the adjustment cost all come under the heading of specific technology; these methods must be worked out at the individual factory.

The quality control techniques seek to reduce the quality control cost without changing the current process, diagnostic methods, and adjustment methods. This is a "soft" technology that can be applied to all production processes. Two methods are described below: one for determining the diagnostic interval n and the other for preventive maintenance by regular replacement of tools.

The optimum diagnostic interval n is given by the formula:

$$n = \sqrt{\frac{2(\bar{u}+l)B}{A-C/\bar{u}}} \qquad\qquad ......(5.8)$$

In the case of the above reaming process, for example:

$$n = \sqrt{\frac{2(2570+1)\times400}{8000-20000/2570}} \fallingdotseq 16 \qquad\qquad ......(5.9)$$

If the diagnostic interval is set at 16 units, the quality control cost L will be:

$$L = \frac{400}{16} + \frac{16+1}{2}\cdot\frac{8000}{2570} + \frac{20000}{2570} + \frac{1\times8000}{2570}$$
$$= 25.0 + 26.5 + 7.8 + 3.1$$
$$= ¥\,62.4 \qquad\qquad ......(5.10)$$

This is the cost per unit, so the saving of 72.4 − 62.4 = ¥ 10 represents an annual saving of ¥ 360,000. The value of L is virtually unchanged when the value of n is altered by 25%. If n = 20, for example:

$$L = \frac{400}{20} + \frac{21}{2}\cdot\frac{8000}{2570} + \frac{20000}{2570} + \frac{8000}{2570} = ¥\,63.6 \qquad\qquad ......(5.11)$$

The difference between this and the value in equation 5.10 is only ¥ 1.2. Accordingly, there can be an error of up to about 30% in the system parameters A, B, C, $\bar{u}$, and l, or after the optimum diagnostic interval n is found, it can be varied within a range of about 25%.

Quality control savings can also be achieved through preventive maintenance. There are two types of preventive maintenance: regular inspection and periodic replacement. In periodic replacement, a part that may cause failure is replaced at periodic intervals without diagnosis. If the mean tool life is 3,000 units, for instance, the tool might be replaced after machining the 2,000th unit without checking whether it is still serviceable.

In regular inspection, the product is checked at regular intervals of n. Even if it is within its specification limits, if there is danger of going outside the specification limits by the next inspection, the tool is replaced at that point. The discussion below is restricted to periodic replacement.

Most of the failures in the reaming process described above stem from tool trouble. Suppose the reamer is replaced periodically every u' = 1,500 units, well before the mean interval between failures of u 2, 570 units. Let C' be the cost per replacement, and suppose that C' is ¥ 18,000, slightly less than the adjustment cost C. Assume that the failure rate before the 1,500th unit is 0.02, which includes failures due to causes other than the reamer itself, such as bending of the reamer by pinholes in the cylinder block. The predicted interval between failures u is now 75,000, calculated as follows:

$$\bar{u} = \frac{1500}{0.02} = 75,000 \text{ units} \qquad \qquad ......(5.12)$$

The new optimum diagnostic interval is therefore:

$$n = \sqrt{\frac{2 \times (75000 + 1) \times 400}{8000 - 20000/75000}} = 87 \doteqdot 100 \text{ units} \qquad ......(5.13)$$

The loss L is now:

L = cost of preventive maintenance
　　+ cost of diagnosis and adjustment

$$= \frac{C'}{u'} + \left(\frac{B}{n} + \frac{n+1}{2} \cdot \frac{A}{\bar{u}} + \frac{C}{\bar{u}} + \frac{lA}{\bar{u}} \right)$$

$$= \frac{18000}{1500} + \left(\frac{400}{100} + \frac{101}{2} \times \frac{8000}{75000} + \frac{20000}{75000} + \frac{1 \times 8000}{75000} \right)$$

$$= 12.0 + 9.8$$
$$= ¥ 21.8 \qquad \qquad ......(5.14)$$

This is an improvement of 63.6 − 21.8 = ¥ 41.8 per unit over the case without preventive maintenance. The annual saving is ¥1.5 million. If each of the 28 steps in the cylinder block manufacturing process could be improved this much, the annual saving would be ¥42 million.

If this quality control improvement is thought of in terms of reducing the failure rate without raising the cost, it is equivalent to increasing the mean interval between failures by a factor of 6.3. Preventive maintenance is thus as valuable as a process-specific technological solution that cuts the failure rate by a factor of 6.3 without increasing the cost. For details, see chapters 4, 5, and 6 in reference (2).

Reference (2) shows that formulas 5.6 and 5.8 can be used as adequate approximations regardless of the distribution of the number of units produced before a failure, and regardless of the percent defective at the time when a failure occurs.

(2) Prediction and Correction: Feed-back Control System Design

If a characteristic has a target value and a signal factor that can be used to correct deviations from it, the signal factor can be used for quantitative control. For example, press pressure can be used as a signal factor to control the thickness of sheet steel, and fuel flow rate can be used as a signal factor to control temperature.

In controlling a quantitative value, the following three problems must be solved.

(a) The measurement interval must be optimized.
(b) From the measured value, the mean characteristic value during the interval up to the next measurement must be predicted.
(c) The optimum amount of correction for differences between the target value and predicted value must be determined.
    If these problems can be solved, then:
(d) The characteristic value can be corrected by changing the level of the signal factor.

In determining the optimum correction interval, periodic analysis (a type of analysis of variance) and the following variability loss formula are useful.

$$L = k\sigma^2 \qquad\qquad ......(5.15)$$

where,
$$k = \frac{\text{loss from not meeting tolerance}}{(\text{tolerance})^2} \qquad ......(5.16)$$

$$\sigma^2 = \text{mean squared deviation from target} \qquad .....(5.17)$$

The simplest method of prediction (b) is to use the measured value itself as the estimated mean for all products in the interval up to the next measurement. There are various more sophisticated methods, but in any case, it is important to find the variance σ_p^2 of the error.

If y is the predicted value and y_0 is the target value, then the optimum amount of correction (c) is given by the following formula:

$$-\beta(y - y_0) \qquad\qquad ......(5.18)$$

where

$$\beta = \begin{cases} 0 & \text{when } F_0 = \dfrac{(y - y_0)^2}{\sigma_p^2} \leq 1 \\[3mm] 1 - \dfrac{1}{F_0} & \text{when } F_0 = \dfrac{(y - y_0)^2}{\sigma_p^2} > 1 \end{cases} \qquad ......(5.19)$$

Recently, more and more production systems are appearing in which steps (a), (b), (c), and (d) above are all performed by an automatic machine or robot. In such systems, on-line quality control focuses on calibration of the sensors (instrumentation) of the automatic machinery or robot, and on diagnosis of the "hunting" phenomenon. Here humans still have to perform steps (a), (b), (c), and (d).

Suppose that the tolerance of an objective characteristic is $\pm\Delta$, that the loss caused by a defective unit is A, that at present the objective characteristic (or a contributing characteristic) is measured at intervals on n_0 units, and that the process is adjusted to the target value when the measured value is off target by $\pm D_0$ or more. Let B be the cost of measurement, l be time lag and C be the cost of adjustment. In a simple system in which the mean squared drift is proportional to the production quantity, the following simplified feed-back control system design procedure can be followed.

(a) Find the average adjustment interval u_0.

$$u_0 = \frac{\text{Total number of products}}{\text{Number of adjustments}} \qquad ......(5.20)$$

(b) Find the optimum measurement interval n and the optimum control limit D.

$$n = \sqrt{\frac{2\,u_0\,B}{A \times D_0^2} \times \Delta} \qquad ......(5.21)$$

$$D = \left(\frac{3\,C}{A} \times \frac{D_0^2}{u_0} \times \Delta^2\right)^{\frac{1}{4}} \qquad ......(5.22)$$

If the optimum solution for n is very different from the current value, set a provisional value between the two and find u_0 again.

(c) Compare the present loss function L_0 with the loss function L for the optimum solution.

$$L_0 = \frac{B}{n_0} + \frac{C}{u_0} + \frac{A}{\Delta^2} \left[\frac{D_0^2}{3} + \left(\frac{n_0+1}{2} + \ell \right) \frac{D_0^2}{u_0} \right] \qquad ......(5.23)$$

$$L = \frac{B}{n} + \frac{C}{u} + \frac{A}{\Delta^2} \left[\frac{D^2}{3} + \left(\frac{n+1}{2} + \ell \right) \times \frac{D^2}{u} \right] \qquad ......(5.24)$$

The parameter u_0 is the current mean adjustment interval. The optimized u can be predicted approximately as $(u_0 D^2/D_0^2)$, but if actual data can be obtained, the actual value should be used.

Note: In control of a process parameter (process condition), if β is the effect on the objective characteristic of a unit change in the parameter, then the calculation can be carried out using Δ_0 in place of Δ:

$$\Delta_0 = \frac{\Delta}{\beta}$$

Problem: The tolerance of a component dimension is $\pm 15 \, \mu m$. The dimension is checked four times a day with time lag 5 and adjusted to target whenever it is found to be outside the control limit of $\pm 2.0 \, \mu m$. The average adjustment is once a half day.

Production volume is 8,000 units a day with five eight-hour days per week. The loss A from a defective unit is ¥60, the measurement cost B is ¥160, the adjustment cost C is ¥40, and at present the adjustment is performed once every two measurements. Find the optimum measurement interval, and calculate the saving over the present practice. Assume that the mean square drift is proportional to the production volume.

Solution: First we find the average adjustment interval u_0:

$$u_0 = \frac{8000}{2} = 4000 \qquad ......(5.25)$$

For this we calculate the optimum measurement interval n and optimum control limit D:

$$n = \sqrt{\frac{2 \, u_0 \, B}{A D_0^2}} \times \Delta$$

$$= \sqrt{\frac{2 \times 4000 \times 160}{60 \times 2.0^2}} \times 15 \fallingdotseq 1000 \qquad ......(5.26)$$

$$D = \left(\frac{3\,C}{A} \times \frac{D_0^2}{u_0} \times \Delta^2 \right)^{\frac{1}{4}}$$

$$= \left(\frac{3 \times 40}{60} \times \frac{2.0^2}{4000} \times 15^2 \right)^{\frac{1}{4}}$$

$$= 0.8 \fallingdotseq 1.0 \qquad \qquad ......(5.27)$$

Then we compare the loss functions. $u_0 = 4,000$ so:

$$u = u_0 \times \frac{D^2}{D_0^2} = 4000 \times \frac{1^2}{2^2} = 1{,}000$$

$$L_0 = \frac{160}{2000} + \frac{40}{4000} + \frac{60}{15^2} \left[\frac{2^2}{3} + \left(\frac{2001}{2} + 5 \right) \frac{2.0^2}{4000} \right] = ¥\,0.71$$
$$......(5.28)$$

$$L = \frac{160}{1000} + \frac{40}{1000} + \frac{60}{15^2} \left[\frac{1.0^2}{3} + \left(\frac{1001}{2} + 5 \right) \times \frac{1^2}{1000} \right] = ¥\,0.42$$
$$......(5.29)$$

There will accordingly be a saving of ¥ 0.29 per production unit, or ¥ 0.58 million per year.

(3) Measurement and Action (Inspection)

Here the action is directed toward the product, whereas in diagnosis and adjustment, and in prediction and correction, it is directed toward the process. The characteristic value of the product is measured and compared with the specification on a pass/fail basis. Units that fail are reworked or scrapped. A simple example will illustrate the method of calculation.

In a certain driving mode, the specification limit for toxic components in automobile exhaust gas is 2.4 g. The cost of inspection using a substitute characteristic is ¥ 800. Automobiles that fail inspection are reworked before shipment, at a cost, including reinspection, of ¥ 12,000. When toxic components were measured on fifteen automobiles, the results were:

1.6, 1.2, 1.5, 2.1, 2.8, 1.7, 2.0, 1.8, 1.4, 1.0, 1.5, 2.5, 2.2, 1.4, 1.8

We shall calculate the profit and loss resulting from 0% inspection and 100% inspection from the above data. (If this calculation is done at the factory, there should be data for at least 50 automobiles.) The loss per unit when there is no (0%) inspection is given by the formula:

$$L_0 = \frac{\text{loss from not meeting tolerance}}{(\text{tolerance})^2} \times \sigma^2 \qquad ......(5.30)$$

Here σ^2 is the mean square of the difference from the target value of zero.

$$\sigma^2 = \frac{1}{15}(1.6^2 + 1.2^2 + \cdots + 1.8^2)$$

$$= 3.34 \qquad\qquad\qquad\qquad ......(5.31)$$

Accordingly,

$$L_0 = \frac{12000}{2.4^2} \times 3.34 = ¥\, 6,958 \qquad\qquad ......(5.32)$$

If 100% inspection is performed with reworking in case of failure, then on the assumption that the mean square of the reworked units will be about the same as the mean square of the units that pass inspection,

$$
\begin{aligned}
L =\ & \text{inspection cost} + (\text{loss from failure in inspection}) \\
& \times \text{percent that fail inspection} \\
& + \text{loss for units that pass inspection}
\end{aligned}
$$

$$= 800 + 12000 \times \frac{2}{15} + \frac{12000}{2.4^2} \times \frac{1}{13}(1.6^2 + 1.2^2 + \cdots + 1.8^2)$$

$$= 800 + 1600 + 2083 \times 2.772$$

$$= ¥\, 8,174 \qquad\qquad\qquad ....(5.33)$$

Even though two of the fifteen units were defective, 100% inspection turns out to be unprofitable. This type of contradiction arises because in general reworking does not guarantee a good unit. This is another reason why quality control should be targeted at the process, by diagnosis and adjustment or prediction and correction. We shall thus forgo further discussion of inspection.

The Roles of Off-Line and On-Line Quality Control

Table 5.5 summarizes the ways to combat variability in functional quality. The symbols in the table have the following meanings:

◎ Possible
○ Possible, but should be a last resort
✕ Impossible

Table 5.5   Means of Functional Quality Countermeasures

Department Countermeasure			Type of noise		
			External	Internal	Unit-to-unit
Off-line quality control	R&D	(1) System design	◎	◎	◎
		(2) Parameter design	◎	◎	◎
		(3) Tolerance design	○	◎	◎
	Production engin-eering	(1) System design	×	×	◎
		(2) Parameter design	×	×	◎
		(3) Tolerance design	×	×	◎
On-line quality control	Production	(1) Process diagnosis and adjustment	×	×	◎
		(2) Prediction and correction	×	×	◎
		(3) Measurement and action	×	×	◎
	Customer relations	After-sales service	×	×	×

As Table 5.5 indicates, by the time the production stage is reached, neither on-line nor off-line measures are effective in combatting internal and external noise. This is why quality problems involving such internal and external noise are called design quality problems. It is extremely important to remember that good product design can solve not only design quality problems but also production quality problems.

It should be noted that R&D as defined in the United States does not include quality control. R&D is the key to both design quality and production quality, but in the United States, notwithstanding the current fad for total quality control, TQC is formally excluded from R&D. This is clearly because of the American concept of quality control and the assumption of linear mathematical models in experimental design.

It is therefore hoped that this book will provide some ideas for a rational theory and practice of quality control, particularly in the field of functional variability.

PROBLEMS

1. The table below gives data (number of non-top-grade tiles per 100) for the tiles fired under inferior conditions on the outside in the experiments described in Table 5.3.

(a) Estimate the factorial effects.
(b) Determine the optimum conditions

No. \ Factor Column	A 1	B 2	C 3	D 4	E 5	F 6	G 7	Number defective per 100 (outside)
1	1	1	1	1	1	1	1	47
2	1	1	1	2	2	2	2	41
3	1	2	2	1	1	2	2	20
4	1	2	2	2	2	1	1	28
5	2	1	2	1	2	1	2	52
6	2	1	2	2	1	2	1	100
7	2	2	1	1	2	2	1	80
8	2	2	1	2	1	1	2	100

2. An injection molding process that produces 12 units per shot is diagnosed every 200 shots. The loss A from a defective unit is ¥ 20, the cost of diagnosis B is ¥ 400 per shot, the failure rate is 1 per 1, 200 shots, and in most cases a failure produces only 1 defective unit. The adjustment cost C when a defective unit is found is ¥8,000, with a time lag 1 of 4 shots.
(a) Using the individual unit (one of the twelve) as the production unit, find the optimum diagnostic interval and calculate the saving as compared with present practice.
(b) Replacing the molds every 800 shots reduced the failure rate by 20%; what was the gain from this regular replacement? Assume that the cost C' of regular replacement is equal to C.

3. The thickness tolerance for an emulsive coating is ±4.0 μm, and a unit variation (1 pois) in the viscosity of the emulsion causes a variation of 0.8 μm in the coating thickness. The loss A when the thickness is out of tolerance is ¥ 800 per square meter. At present the viscosity is checked once every two hours, at a cost B of ¥ 400 with a time lag 0.
     Viscosity is controlled by controlling the amount of solvent. The adjustment cost C is ¥ 1,200. At present the control limit is ±1.0, and control action has been found necessary in one of three measurements. Design an optimum measurement and correction system. The production volume is 50,000 m² per hour.

4. An aircraft component is checked every flight (n = 6 hours) at a cost B of ¥ 2,000 per check. A problem is found and the component replaced about once every 100 checks. The cost C* of the replacement is ¥ 60,000. If the component does not function

properly and causes an accident when other primary trouble occurs during that time, the estimated cost is ¥ 15 billion. Such malfunctions occur about three times a year. About 5,000 hours of flight time are logged per year, and average flight duration is 6 hours. Find the optimum checking interval, and calculate the improvement over present practice. The per-hour loss function L is given by the formula:

$$L = \frac{B}{n} + \frac{C^*}{u^*} + \frac{n+1}{2} \times \frac{1}{u^*} \times \frac{C}{u_0}$$

where u^* is the mean maintenance interval, and u_0 is the interval between primary troubles.

DISCUSSION

Design and Modification

S (Student):   Let's go back to the dyer who is shown a color sample and uses his professional skill to match it with a test dye. The test succeeds if the result matches the sample. If it differs in hue or saturation, the dyer changes the mixture of dyes or the amount per weight of cloth until he gets the target value. Is this really that much different from design work?

G (Genichi):   It's certainly not the kind of design work that should be done in an R&D department. You could call it a form of production design in the broad sense, but it is closer to modification or calibration.

S: But even R&D people have to adjust parameters to target values.

G: As the text pointed out, in adjusting to a target value, if there are three target characteristics, all you need are three factors with substantial impact on them. You don't have to study all the factors with large effects.

S: But in studying color balance in television or for film, for example, you couldn't experiment with all possible saturations of all the colors.

G: In that case, you could fit the target curve by taking three points (low, medium, and high intensity) for each of the three primary colors. It might not be easy to get an exact fit to the target curve, but if you could get five points on it, you would have a pretty good match. Five high- impact factors with a large influence should be sufficient.

S: So matching a target value is just an adjustment process, not part of design research.

G: But some targets are quite difficult to match. Take 100% thermal efficiency, for example, or reducing noise or toxic exhaust gas to 1/100th of their present values. For hue, saturation, dimension, hardness, voltage, clearance, air speed, and so on, we're talking about cases in which there is a signal factor that can be used to adjust the characteristic to an arbitrary value. Since the signal factor is used to correct deviations from target, it should have a strong, linear effect.

S: For hue and saturation, there are bound to be controllable factors, so if there are four effective signal factors, I guess you can achieve the target value. All you need to do is to find the relationships and solve the corresponding simultaneous equations.

G: If the impact is large enough and that if you're solving four simultaneous linear equations, the accuracy of the solution is also a problem, but the accuracy can be found by experimenting with the four factors, using an orthogonal array.

S: With four objective characteristics, all we need are the separate effects of four factors on them. But what if the equations are nonlinear?

G: For a signal factor, it's best if only the linear term has substantial size. Accordingly, you should examine a large number of factors and select only those that have large linear effects. This question, however, is related to the question of signal-to-noise ratio, which will be covered in Chapter 10. I referred to adjustment as production technology, but finding the most efficient factor to adjust is the work of the R&D department.

S: As is the study of dynamic characteristics.

6 PARAMETER DESIGN AND TOLERANCE DESIGN: CASE STUDY

Setting Parameter Levels

Methods of determining tolerances were explained in Chapters 2 to 4. The most important aspect of design work (both product and process design), however, is the setting of the parameter levels (mid-values and specifications) after the system has been decided on. Although tolerances can improve function and quality, they add cost. They are a necessary evil that should be avoided as far as technically possible.

Although this discussion focuses on product design, it applies equally well to production process design, which is the same thing if one thinks of the process as a product that produces manufactured goods. As noted in the preceding chapter, there are three steps in product design.

(1) System design (functional design). The purpose of this step is to design a product system that possesses the functions designated by the planning department. The designer may select a single system or may choose to work with two or three candidate systems. There are many design systems with the same functions, but there are no rules for choosing which is best. Accordingly, system design frequently relies upon on judgment.

(2) Parameter design. After the system is decided on, the purpose of parameter design is to select the optimum levels for the system parameters. This is the most important part of the design process, because it can both improve quality and lower cost.

If the system parameters vary, the objective (output) characteristic *y* will deviate from target, causing a loss. Deviation in the characteristic value *y* is caused not only by system parameter variation but also by variation in environmental conditions. One can try to control the deviation of the target characteristic *y* by setting tight tolerances for the initial values and rates of deterioration of the system components and materials -- i.e., by tolerance design for the system parameters -- but this increases the cost since it requires high- quality components and materials, and it does not deal with environmental variations.

The purpose of parameter design is to adjust the parameter levels (mid- values and specifications) so that the objective characteristic will not vary much even if the system and environmental parameters change. It is a search for the parameter levels at which the characteristic *y* is stable despite the use of inexpensive components and materials. The most important precondition for designing an inexpensive but good product is that it be inexpensive. Starting with inexpensive components and reducing the variability about the mean of the objective characteristic *y* without raising the cost is an advanced design technique also known as the use of nonlinearity.

This technique leaves the error factors that affect the output characteristic unchanged, or even lets them vary more than before, but still keeps the output characteristic steady. Accordingly, it makes use of interaction with all the error factors. Overall parameter design will be covered in the later chapters. Here we will discuss the method in connection with the Wheatstone bridge, a textbook circuit that in this case is intended to measure the resistance of a product that has a nominal value of two ohms.

Parameter Design of Wheatstone Bridge

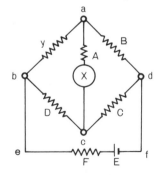

Figure 6.1 Wheatstone Bridge and Parameter Symbols

The problem with the Wheatstone bridge is how to set the nominal values (not the tolerances) of the parameters *A, C, D, E,* and *F* so that the unknown resistance y can be measured accurately. *A, C, D, E*, and *F* are the controllable factors. The resistance *B* is not controllable because it is adjusted during the measurement to make the galvanometer x read zero.

In Table 6.1, three levels are set for each of the controllable factors *A, C, D, E*, and *F*. Level 1 is one-fifth of level 2, which is one-fifth of level 3.

Table 6.1 Levels for Controllable Factors

Factor	Level 1	Level 2	Level 3
A (ohms)	20	100	500
C (ohms)	2	10	50
D (ohms)	2	10	50
E (volts)	1.2	6	30
F (ohms)	2	10	50

The purpose of parameter design is to evaluate the overall variation due to internal and external noises for different levels of the controllable factors, and to find a design that is as immune as possible to noise effects. Accordingly, the designer must know what types of internal and external noise are present, and, at least for the major noise factors, must set levels and investigate the size of their effect.

Among the noise factors in this problem are variability in the bridge components. Starting with inexpensive components and estimating possible error in their characteristic values, we can set three error levels for each factor as shown in Table 6.2. Another error factor is the error in reading the galvanometer X. Assuming that when the galvanometer is read as zero, there may actually be a current of about 0.2 milliamperes, we can again set three levels.

Table 6.2 Levels of Error Factors

Factor	Level 1	Level 2	Level 3
A' (%)	−0.3	0	0.3
B' (%)	−0.3	0	0.3
C' (%)	−0.3	0	0.3
D' (%)	−0.3	0	0.3
E' (%)	−5.0	0	5.0
F' (%)	−0.3	0	0.3
X' (mA)	−0.2	0	0.2

The error in the power supply voltage has practically no effect on the measurement error, so we can use the most inexpensive battery on the market and set error levels of -5%, 0, and +5% from the mid-value. Whether the aim is to design an accurate measurement system or a good, stable product, one should always start with components and materials of the least expensive grade, or at most a medium grade, and determine their influence by setting appropriate levels. The reason for this is that a design strategy to improve accuracy, stability, and reliability should start with a parameter design study. Tolerance design should be reserved for dealing with particularly stubborn error factors after parameter design is completed. In the parameter design study, the effect of the error factors can be best found by letting the factors vary widely, so it would be counterproductive to conduct the study with expensive, high-quality components and materials.

A Wheatstone bridge is used to measure an unknown resistance y, which is connected across the two points a and b as shown in Figure 6.1. The resistor B is adjusted until the current X registered by the galvanometer is zero, at which point the resistance value B is read and y is calculated from the formula:

$$y = \frac{BD}{C}$$

......(6.1)

Since we are studying the measurement error, however, we will suppose that the current is not exactly zero, but that a positive or negative current of about 0.2 mA is flowing. In that case the resistance is not the value in equation 6.1. A well-known approximation to the true value is given by the following formula (proof omitted).

$$y = \frac{BD}{C} - \frac{X}{C^2 E} [A(D+C) + D(B+C)] [B(C+D) + F(B+C)]$$

......(6.2)

The three levels of the five controllable factors A, C, D, E, and F are assigned to columns 1, 3, 4, 5, and 6 of the L_{36} orthogonal array. An orthogonal array to which controllable factors are assigned is called an inner array. The L_{36} array is the array most frequently used in parameter and tolerance design work, because (1) there are usually three levels set for both the controllable factors and the error factors, and (2) the interactions between controllable factors, or between error factors, are usually not calculated, so they must be treated as error. It is thus desirable to have the effects of the interactions distributed as evenly as possible, and the L_{36} array distributes the interaction of any two columns nearly uniformly across all the columns.

The controllable factors are, of course, often assigned to other orthogonal arrays. Also, if the data are collected by doing actual experiments, the L_{36} array requires considerable time and expense, so various sophisticated layout techniques have been developed. See reference (1).

Layout and Signal-to-Noise Ratio

In the inner L_{36} orthogonal array shown in the lower left of Table 6.3, the 36 experiments correspond to 36 combinations of the mid-values of the controllable factors. The error factors (levels of deviation from the nominal values) in Table 6.2 are assigned to a separate L_{36} orthogonal array, called an outer array, shown in the upper right of Table 6.3. The entire outer L_{36} array is applied to each of the 36 experiments in the inner array, so there are $36 \times 36 = 1,296$ combinations in all for which to calculate the value of y from equation 6.2 and find the difference from the true value. The layout in Table 6.3 is called the direct product layout of the inner and outer L_{36} orthogonal arrays.

The error factors A', B', C', D', E', F', and X' are assigned to columns 1, 2, 3, 4, 5, 6, and 7 of the outer L_{36} array. The levels in the outer array corresponding to experiment No. 2 of the inner array are given in Table 6.4.

Table 6.3 Direct Product Layout

Col \ No.						1	2	3	4	⋯	36
A'	1					1	2	3	1	⋯	3
B'	2					1	2	3	1	⋯	2
C'	3					1	2	3	1	⋯	3
⋮	⋮					⋮	⋮	⋮	⋮	⋱	⋮
	13					1	1	1	1	⋯	3

Col \ No.	A	e	C	⋯	e						
	1	2	3	⋯	13	13					
1	1	1	1	⋯	1	y_1 y_2 y_3 y_4 ⋯ y_{36}					
2	2	2	2	⋯	1	y_1 y_2 y_3 y_4 ⋯ y_{36}					
3	3	3	3	⋯	1	y_1 y_2 y_3 y_4 ⋯ y_{36}					
4	1	1	1	⋯	1	y_1 y_2 y_3 y_4 ⋯ y_{36}					
⋮	⋮	⋮	⋮	⋱	⋮	⋮ ⋮ ⋮ ⋮ ⋮ ⋮					
36	3	2	3	⋯	3	y_1 y_2 y_3 y_4 ⋯ y_{36}					

In Table 6.5, the column marked "Experiment No. 2" shows how the measurement error of the unknown resistance y varies for the 36 combinations of error factors in the outer L_{36} array. All these data are for the levels of the controllable factors in experiment No. 2 in the inner array ($A_2C_2D_2E_2F_2$). The mid-values are 100 ohms for resistance A, 10 ohms for resistances C and D, 6 volts for the power supply E, and 10 ohms for resistance F. In the outer array, the

Introduction to Quality Engineering

resistances vary by ±0.3% around the nominal values, and the supply voltage E by ±5%, giving the three levels of the error factors A', C', D', E', and F' listed in Table 6.4. For resistance B and galvanometer reading X, the nominal values are always 2 ohms and 0 amperes, and the three levels of B' and X' are as shown in Table 6.4.

Teble 6.4 Levels of Error Factors for Experiment No. 2 in the Inner Array

Factor	1	2	3
A' (Ω)	99.7	100.0	100.3
B' (Ω)	1.994	2.0	2.006
C' (Ω)	9.97	10.0	10.03
D' (Ω)	9.97	10.0	10.03
E' (V)	5.7	6.0	6.3
F' (Ω)	9.97	10.0	10.03
X' (A)	−0.0002	0	0.0002

The measurement error data for experiment No. 2 in Table 6.5 were obtained by calculating the resistance reading y at the error factor levels in columns 1 to 7 and subtracting 2 ohms. The sum of the squares of the 36 measurement errors, denoted S_T, is calculated as follows.

$$S_T = 0.1123^2 + 0.000^2 + \ldots\ldots + (-0.0120)^2 \qquad \ldots\ldots(6.3)$$
$$= 0.31141292 \ (f = 36) \qquad \ldots\ldots(6.4)$$

Actually, the four-place figures in equation 6.3 give 0.31140717. The slightly different value in 6.4 was obtained using a computer at a higher level of precision.

The error variance V_T can now be found as follows.

$$V_T = \frac{S_T}{36} = \frac{0.31141292}{36} = 0.00865036 \qquad \ldots(6.5)$$

This is the variance of the measurement error when the error factors range in size as shown in Table 6.2 and all the controllable factors are at level 2, under the conditions of experiment No. 2. If the combination of controllable factor levels is changed and the error factors are again varied as in Table 6.2, the variance of the measurement error will change. The variation of error S_e and the variation of mean S_m for all 36 experiments in the inner orthogonal array are given by the formulas:

$$S_e = \text{(total squared error)} - \text{(correction factor)} \qquad \ldots\ldots(6.6)$$

Table 6.5 Layout of Error Factors and Measurement Error Data

No.	A' 1	B' 2	C' 3	D' 4	E' 5	F' 6	X' 7	8	9	10	11	12	13	(1) Experiment No. 2	(2) Optimum conditions
1	1	1	1	1	1	1	1	1	1	1	1	1	1	0.1123	−0.0024
2	2	2	2	2	2	2	2	2	2	2	2	2	1	0.0000	0.0000
3	3	3	3	3	3	3	3	3	3	3	3	3	1	−0.1023	0.0027
4	1	1	1	1	2	2	2	2	3	3	3	3	1	−0.0060	−0.0060
5	2	2	2	2	3	3	3	3	1	1	1	1	1	−0.1079	−0.0033
6	3	3	3	3	1	1	1	1	2	2	2	2	1	0.1252	0.0097
7	1	1	2	3	1	2	3	3	1	2	2	3	1	−0.1188	−0.0036
8	2	2	3	1	2	3	1	1	2	3	3	1	1	0.1009	−0.0085
9	3	3	1	2	3	1	2	2	3	1	1	2	1	0.0120	0.0120
10	1	1	3	2	1	3	2	3	2	1	3	2	1	−0.0120	−0.0120
11	2	2	1	3	2	1	3	1	3	2	1	3	1	−0.1012	0.0086
12	3	3	2	1	3	2	1	2	1	3	2	1	1	0.1079	0.0033
13	1	2	3	1	3	2	1	3	3	2	1	2	2	0.0950	−0.0087
14	2	3	1	2	1	3	2	1	1	3	2	3	2	0.0120	0.0120
15	3	1	2	3	2	1	3	2	2	1	3	1	2	−0.1132	−0.0035
16	1	2	3	2	1	1	3	2	3	3	2	1	2	−0.1241	−0.0096
17	2	3	1	3	2	2	1	3	1	1	3	2	2	0.1317	0.0215
18	3	1	2	1	3	3	2	1	2	2	1	3	2	−0.0120	−0.0120
19	1	2	1	3	3	3	1	2	2	1	2	3	2	0.1201	0.0153
20	2	3	2	1	1	1	2	3	3	2	3	1	2	0.0000	0.0000
21	3	1	3	2	2	2	3	1	1	3	1	2	2	−0.1250	−0.0154
22	1	2	2	3	3	1	2	1	1	3	3	2	2	0.0060	0.0060
23	2	3	3	1	1	2	3	2	2	1	1	3	2	−0.1247	−0.0096
24	3	1	1	2	2	3	1	3	3	2	2	1	2	0.1138	0.0035
25	1	3	2	1	2	3	3	1	3	1	2	2	3	−0.1129	−0.0035
26	2	1	3	2	3	1	1	2	1	2	3	3	3	0.0951	−0.0087
27	3	2	1	3	1	2	2	3	2	3	1	1	3	0.0120	0.0120
28	1	3	2	2	2	1	1	3	2	3	1	3	3	0.1186	0.0095
29	2	1	3	3	3	2	2	1	3	1	2	1	3	−0.0060	−0.0060
30	3	2	1	1	1	3	3	2	1	2	3	2	3	−0.1197	−0.0036
31	1	3	3	3	2	3	2	2	1	2	1	1	3	0.0060	0.0060
32	2	1	1	1	3	1	3	3	2	3	2	2	3	−0.1133	−0.0093
33	3	2	2	2	1	2	1	1	3	1	3	3	3	0.1194	0.0036
34	1	3	1	2	3	2	3	1	2	2	3	1	3	−0.0957	0.0087
35	2	1	2	3	1	3	1	2	3	3	1	2	3	0.1194	0.0036
36	3	2	3	1	2	1	2	3	1	1	2	3	3	−0.0120	−0.0120

Where correction factor $= \dfrac{(total)^2}{36}$

$$S_m = \frac{(2 \times 36 + total)^2}{36} \qquad \qquad ......(6.7)$$

When these have been calculated, the signal-to-noise (S/N) ratio η, discussed elsewhere in this chapter, can be calculated:

$$\eta = \frac{\frac{1}{36}(S_m - V_e)}{V_e} \qquad \qquad ......(6.8)$$

Where V_e is error variance.

$$......(6.9)$$

$$V_e = \frac{S_e}{35}$$

The S/N ratio gives an estimate of the following ratio:

$$\frac{\text{Numerator}}{\text{Denominator}} = \frac{m^2}{o^2} \qquad \qquad ......(6.10)$$

This is because S_m is an estimate of $o^2 + 36m^2$. The reason for using the S/N ratio instead of the error variance will not be discussed in detail here, but, briefly, it is because in this measurement system the tendency for the measured resistance value to be too high or too low can be calibrated using a standard resistor. Even if that were not the case, there would be no problem in using the S/N ratio, because for any set of conditions in the inner array, if the mean estimated true value is equal to the true value of m, the numerator m^2 is a constant and the result is equivalent to comparing the error variance alone.

S/N Analysis

Table 6.6 gives the S/N ratio as found from formula 6.8 for each of the 36 experiments in the inner orthogonal array. The data in Table 6.6 are the common logarithms of the S/N ratios multiplied by 10.

The S/N ratio is the reciprocal of the variance of the measurement error, so it is maximal for the combination of parameter levels that has the minimum error variance. This is the optimum design. Accordingly, the data for the factors assigned to the inner orthogonal array can be analyzed by using the decibel value as the objective characteristic and making the usual level-to-level comparison. For example, the main effect of A is:

$$S_A = \frac{378.7^2 + 225.4^2 + 80.9^2}{12} - \frac{685.0^2}{36} \qquad \dots\dots(6.11)$$

$$= 3700.21 \quad (f = 2) \qquad \dots\dots(6.12)$$

S_C, S_D, S_E, and S_F can be found in the same way. The total variation S_T is calculated as:

$$S_T = 32.2^2 + 26.7 + \dots + 8.0^2 - \frac{685.0^2}{36} \qquad \dots\dots(6.13)$$

$$= 11397.42 \quad (f = 35) \qquad \dots\dots(6.14)$$

Table 6.6 Layout of Controllable Factors (Inner Array) and S/N Ratios

No.	A 1	e 2	C 3	D 4	E 5	F 6	e 7	e 8	e 9	e 10	e 11	e 12	e 13	S/N ratio (decibels)
1	1	1	1	1	1	1	1	1	1	1	1	1	1	32.2
2	2	2	2	2	2	2	2	2	2	2	2	2	1	26.7
3	3	3	3	3	3	3	3	3	3	3	3	3	1	15.9
4	1	1	1	1	2	2	2	2	3	3	3	3	1	36.4
5	2	2	2	2	3	3	3	3	1	1	1	1	1	28.6
6	3	3	3	3	1	1	1	1	2	2	2	2	1	7.2
7	1	1	2	3	1	2	3	3	1	2	2	3	1	16.5
8	2	2	3	1	2	3	1	1	2	3	3	1	1	13.0
9	3	3	1	2	3	1	2	2	3	1	1	2	1	28.0
10	1	1	3	2	1	3	2	3	2	1	3	2	1	15.0
11	2	2	1	3	2	1	3	1	3	2	1	3	1	16.4
12	3	3	2	1	3	2	1	2	1	3	2	1	1	25.5
13	1	2	3	1	3	2	1	3	3	2	1	2	2	43.8
14	2	3	1	2	1	3	2	1	1	3	2	3	2	− 8.3
15	3	1	2	3	2	1	3	2	2	1	3	1	2	14.6
16	1	2	3	2	1	1	3	2	3	3	2	1	2	29.0
17	2	3	1	3	2	2	1	3	1	1	3	2	2	6.9
18	3	1	2	1	3	3	2	1	2	2	1	3	2	14.7
19	1	2	1	3	3	3	1	2	2	1	2	3	2	21.5
20	2	3	2	1	1	1	2	3	3	2	3	1	2	17.4
21	3	1	3	2	2	2	3	1	1	3	1	2	2	14.0
22	1	2	2	3	3	1	2	1	1	3	3	2	2	46.5
23	2	3	3	1	1	2	3	2	2	1	1	3	2	5.5
24	3	1	1	2	2	3	1	3	3	2	2	1	2	− 8.2
25	1	3	2	1	2	3	3	1	3	1	2	2	3	27.3
26	2	1	3	2	3	1	1	2	1	2	3	3	3	43.4
27	3	2	1	3	1	2	2	3	2	3	1	1	3	20.9
28	1	3	2	2	2	1	1	3	2	3	1	3	3	44.1
29	2	1	3	3	3	2	2	1	3	1	2	1	3	39.3
30	3	2	1	1	1	3	3	2	1	2	3	2	3	−17.0
31	1	3	3	3	2	3	2	2	1	2	1	1	3	23.0
32	2	1	1	1	3	1	3	3	2	3	2	2	3	44.2
33	3	2	2	2	1	2	1	1	3	1	3	3	3	− 0.9
34	1	3	1	2	3	2	3	1	2	2	3	1	3	43.4
35	2	1	2	3	1	3	1	2	3	3	1	2	3	− 7.7
36	3	2	3	1	2	1	2	3	1	1	2	3	3	8.0

Table 6.7 is an analysis of variance for this example. The data in Table 6.7 were obtained on a computer at a higher level of precision, so the values in 6.12 and 6.14 differ slightly from the exact values of equations 6.11 and 6.13.

Table 6.7 Analysis of Variance of S/N Ratio

Source	f	S	V
A	2	3700.21	1850.10
C	2	359.94	179.97
D	2	302.40	151.20
E	2	4453.31	2226.65
F	2	1901.56	950.78
e	25	680.00	27.20
T	35	11397.42	

In tolerance design, the question was the contribution ratios of the error factors, and it was not necessary to know the estimated values of the resistance for each level of the error factors. When the S/N ratio is the objective characteristic, however, it is important to see how the value varies at different levels of the controllable factors. Determining the significance of the factor effects is a nearly pointless exercise. The combination that gives the better S/N ratio should be selected, even when the difference is slight. If there is a cost difference, the correct choice is the choice that comes out better after the cost difference is subtracted. The average S/N ratios for the different levels of the controllable factors are given in Table 6.8.

Table 6.8 Estimates of Significant Controllable Factors; Average Values on Each Level

(Unit: decibels)

Level	A	C	D	E	F
1	31.56	14.56	20.91	5.66	27.58
2	18.78	21.10	21.24	18.52	19.68
3	6.73	21.42	14.93	32.89	9.81

Figure 6.2 graphs these effects. The confidence limits for the error are also shown in Figure 6.2, although they are not very important.

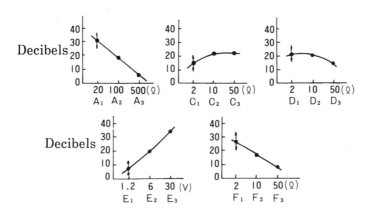

Figure 6.2 Graphs of Significant Factors

The confidence limits in the graphs in Figure 6.2 were found as follows.

$$\pm \sqrt{\frac{F \times V_e}{n_e}}$$

$$\pm \sqrt{\frac{4.24 \times 27.20}{12}} = \pm 3.10$$

where 4.24 is the $F(1,25)$ ratio from F Table,
and n_e = total no. of unit in one level.

From Figure 6.2, the optimum conditions, meaning the conditions which give the largest S/N ratio, are A_1, C_3, D_2, E_3, and F_1. The expected gain from that combination in comparison with the initial combination $A_2 C_2 D_2 E_2 F_2$ is:

$$\begin{aligned}
\text{Gain} &= (31.56 - 18.78) + (21.42 - 21.10) + (21.24 - 21.24) \\
&\quad + (32.89 - 18.52) + (27.58 - 19.68) = 12.78 + 0.32 + 0 \\
&\quad + 14.37 + 7.90 = 35.37 \pm 14.54 \text{ (db)} \qquad \text{......(6.15)}
\end{aligned}$$

The confidence limits for the expected gain are calculated as:

$$\pm \sqrt{F \times V_e \left(\frac{10}{n_e} + 1\right)} = \pm \sqrt{4.24 \times 27.20 \left(\frac{10}{12} + 1\right)} = \pm 14.54$$

We shall confirm the optimum design by calculating the size of the error. Three levels for the error factors A', B', C', D', E', F', and X' are set as in Table 6.2 and assigned to the outer L_{36} orthogonal array. The resulting measurement errors are shown in the column headed "(2) Optimum conditions" in Table 6.5. The sum of the squares of the errors

S_T is:

$$S_T = 0.0024^2 + 0.0000^2 + \cdots + (-0.0120)^2 = 0.00289623$$
......(6.16)

The error variance, including the general mean, is:

$$V_T = \frac{0.00289623}{36} = 0.00008045$$
......(6.17)

This is only 1/107.5 the size of the error variance in equation 6.5, a gain of 20.32 decibels. The improvement is slightly below the lower confidence limit in equation 6.15, but it is more improvement than would be obtained by a tenfold reduction in the tolerances, and it is obtained without resort to high-quality components. This is the benefit of parameter design. Despite the use of low- priced components and materials with high variability, stability can be improved to yield a major reduction in measurement error, or a major improvement in quality in the case of product.

Tolerance Design

In parameter˙ design, the optimum combination of levels of the control factors is chosen. When parameter design cannot sufficiently reduce the effect of internal and external noise, it becomes necessary to restrict the variation of the major noise factors to within narrower ranges, even though this raises the cost. This process is tolerance design. Since the effect of the error factors can often be substantially reduced by parameter design, it is important to perform parameter design first and tolerance design second.

Let us assume that parameter design has been performed as described above. Ordinarily, there should be a follow-up parameter design in which the range of levels is further widened and new controllable factors are introduced, but we shall assume that the optimum conditions have been selected as $A_1 \, C_3 \, D_2 \, E_3 \, F_1$. Still assuming low-cost components, error factors with three levels are set up around the optimum conditions, these are assigned to the outer orthogonal array, and the measurement error is found. It is customary at this stage to study error factors that were not studied in parameter design, but here we shall use the same error factors (A', B', C', D', E', F', and X') and error levels as in Table 6.2. Ideally, the levels for each error factor should be set as follows. Let the nominal value of the error factor be m and its standard deviation be o. The ideal levels are then:

Level 1 $m - \sqrt{\dfrac{3}{2}}\,\sigma$

Level 2 m

Level 3 $m + \sqrt{\dfrac{3}{2}}\,\sigma$

......(6.18)

We shall assume that the levels in Table 6.2 were set in this way. Then the results when the error factors are assigned to the L_{36} orthogonal array and the measurement error is calculated for 36 sets of conditions are as given in the column headed "(2) Optimum conditions" in Table 6.5. For general products other than instruments, the data would represent departure from the target value.

The data in Table 6.5 are multiplied by 10,000, variance analysis is performed, and the contribution levels are calculated. The main effects of the error factors are resolved into linear and quadratic components using orthogonal polynomials.

$$S_m = \frac{4^2}{36} = 0 \quad (f = 1)$$

......(6.19)

$$S_{A_l'} = \frac{(-A'_1 + A'_3)^2}{r \times \lambda^2 S} = \frac{[-(-2) + 3]^2}{12 \times 2} = 1 \quad (f = 1)$$

......(6.20)

$$S_{A_q'} = \frac{(-2 - 2 \times 3 + 3)}{12 \times 6} = 0 \quad (f = 1)$$

......(6.21)

Similar calculations are performed for the other factors, giving the analysis of variance in Table 6.9.

The analysis of variance in Table 6.9 can be simplified by retaining only the four large linear effects B_1, C_1, D_1, and X_1 and pooling the rest with error, as in Table 6.10.

If special-grade resistors with a standard deviation one order of magnitude smaller are used for the high-impact factors B_l, C_l, and D_l and a first-grade galvanometer with a five-times-better reading accuracy is used, the contribution levels for B_l, C_l, and D_l are reduced to 1/100 of their previous values and the contribution level of X_l is reduced to 1/25. The error variance V_e is then given by the formula:

$$V_e = V_T \left[\left(\frac{1}{10}\right)^2 (\rho_{Bl} + \rho_{Cl} + \rho_{Dl}) + \left(\frac{1}{5}\right)^2 \rho_{Xl} + \rho_e \right]$$

$$= 8045.1\,[0.01\,(0.29860 + 0.30074 + 0.30093) + 0.04$$
$$\times 0.09956 + 0.00017]$$

$$= 8045.1 \times 0.013155 = 105.8$$

......(6.22)

Table 6.9 Analysis of Variance for Tolerance Design

Source		f	S	V
m		1	0	0
A	l	1	1	1
	q	1	0	0
B	l	1	86482	86482
	q	1	1	1
C	l	1	87102	87102
	q	1	0	0
D	l	1	87159	87159
	q	1	0	0
E	l	1	0	0
	q	1	0	0
F	l	1	0	0
	q	1	1	1
X	l	1	28836	28836
	q	1	0	0
e		21	41	1.95
T		36	289623	

Table 6.10 Summarized Analysis of Variance

Source	f	S	V	$\rho(\%)$
B_l	1	86482	86482	29.860
C_l	1	87102	87102	30.074
D_l	1	87159	87159	30.093
X_l	1	28836	28836	9.956
e	32	44	1.38	0.017
T	36	289623	8045.1	100.000

Returning to original units, the variance σ^2 of the measurement error is then:

$$\sigma^2 = 0.000001058 \qquad \qquad ......(6.23)$$

Three standard deviations are now very small and $\pm 3\sigma$ are:

$$\pm 3\sigma = \pm 3 \times \sqrt{0.000001058} = \pm 0.0031 \qquad ......(6.24)$$

Summary of Quality Design

In the problems in this chapter, the nominal values the designer first considered for the parameters are shown in column (1) of Table 6.11. The optimum nominal values found by varying the first values by factors of 0.2 and 5 are listed in column (2). The error variance was reduced to less than 1/100 of its original value, still using the same inexpensive grade of components.

Table 6.11   Initial and Optimum Nominal Values
of Parameters, and Error Variance

	(1) Initial nominal value	(2) Optimum nominal value
$A(\Omega)$	100	20
$C(\Omega)$	10	50
$D(\Omega)$	10	10
$E(V)$	6	30
$F(\Omega)$	10	2
Error variance	0.00865036	0.00008045
3σ	±0.2790	±0.0269

Since the aim of parameter design is to improve measurement accuracy while using inexpensive, highly variable components, its methods can be applied in almost all actual measurement situations. Tolerance design, in contrast, seeks to deal with major factors by using expensive components with small variability, so it is necessary to calculate whether the reduction in variability is worth the additional cost in each situation.

In the previous section using special-grade resistors for the three especially large factors A, C, and D, and a first-grade galvanometer was proposed. Assuming that the tolerance specification of the resistance value of the product is 2 ± 0.5 ohms and the loss A when it is out of specification is ¥ 300, we shall calculate the loss for three cases: the original conditions (column (2) in Table 6.5), the optimum conditions after parameter design, and the conditions after tolerance design. We shall assume that parameter design does not alter the cost but that tolerance design raises the cost of each resistor by ¥ 400 and the cost of the galvanometer by ¥ 20,000, and that 50% of these cost increases are in the form of interest and depreciation.

Under the original conditions, the loss due to measurement error is:

$$L = \frac{300}{0.5^2} \times \sigma^2 = 1200 \times 0.00865036 = ¥\,10.3804$$

......(6.25)

Similarly, after parameter design:

$$L = 1200 \times 0.00008045 = ¥\,0.0965$$

...(6.26)

If the tolerance design specifies only the use of special-grade resistors, the loss is:

$$L = \frac{400 \times 0.5 \times 3}{120000} + 1200 \times 0.00000875 = ¥\,0.0155$$

......(6.27)

If the tolerance design specifies both special-grade resistors and a first-grade galvanometer, the loss is:

$$L = \frac{400 \times 0.5 \times 3 + 20000 \times 0.5}{120000} + 1200 \times 0.000001058$$

$$= ¥\,0.0896$$

......(6.28)

The optimum solution is given by equation 6.27. These results are summarized in Table 6.12.

Table 6.12 Quality and Cost at Different Design Stages

Design stage	σ_{out}^2	Quality L (¥)	Cost (¥)	Total (¥)	Annual (¥ '000)
System design only	0.00865036	10.3804	Base	10.3804	1,245.6
After parameter design	0.00008045	0.0965	+0	0.0965	1.16
Special-grade resistors only	0.00000875	0.0105	+0.0050	0.0155	0.19
First-grade galvanometer	0.00000106	0.0013	+0.0883	0.0896	1.08

PROBLEM

When alternating current with a voltage of V volts and frequency of f hertz is applied to a circuit with a resistance R and self-inductance L, the resulting current y in amperes is given by the formula:

$$y = \frac{V}{\sqrt{R^2 + (2\pi f L)^2}}$$

The voltage V has a standard deviation σ_V of 10 volts around a mid-value of 100 volts, and the frequency f varies between the two levels of 50 and 60 hertz. There are three grades of resistors and coils, with the following degrees of variability.

Grade 1: Standard deviation = 1% of mid-value
Grade 2: Standard deviation = 5% of mid-value
Grade 3: Standard deviation = 20% of mid-value

The output current is outside of its specification of 10.0 ± 5.0 amperes, which means that it fails to function properly and causes a loss of ¥ 8,000. Compared with grade-3 components, a grade-1 resistor is ¥ 300 more expensive and a grade-1 coil is ¥ 800 more expensive, while a grade-2 resistor is ¥ 10 more expensive and a grade-2 coil ¥ 20 more expensive. The annual production volume is 500,000 units. Perform parameter design by compounding the error factors into one compound noise factor and tolerance design by assigning the error factors to an L_{18} array.

DISCUSSION

Role of Design

G (Genichi):The purpose of product planning is to estimate what functions and what price would lead to what level of demand, and from that to select the functions and price of the product to be developed. The designer's job is to design the product so that a satisfactory profit can be made by selling it at that price. For many products, the design work breaks down into five steps:
  (1) Total system design
  (2) Subsystem design
  (3) Unit and part design
  (4) Component development
  (5) Material development

At a parts factory, of course, steps (1) and (2) would not be needed. It would only be necessary to develop steps (3), (4), and (5) to meet the customer's specifications. As we have already seen, each of the design and development steps involves three substeps:
  (a) System design
  (b) Parameter design
  (c) Tolerance design

S (Student):Of these three substeps in the development of basic system elements, parameter design is the substep where the specifications of the raw materials, the process conditions, and other conditions are set. The temperature at which to conduct a process, for instance, would be a parameter design question. After this is decided, the range within which to allow the temperature to vary would be a tolerance design question.

G: Yes, and system design would be the study of candidate manufacturing systems -- deciding how to process the raw materials to get components with the required functions, without regard to reliability. Then the target values for the system parameters would be set in parameter design.

S: The required functions of the basic elements are given by the designer of the unit or component. But the total system designer has to decide what subsystems to use and how to integrate them, and has to set target values for the required subsystem characteristics. There might be many different systems with the functions decided on by product planning. The total system designer has to choose among them on the basis of his own knowledge.

G: Actually, you could say the same for all of the five design steps; for a given set of output characteristics, there would be many systems with the desired functions. In this chapter, for instance, the desired function was to measure the value of an unknown resistance y, so any circuit in which an output characteristic varies depending on a resistance value would be a candidate system. The number of candidate systems would be practically infinite.

S: The question is which of all those systems to develop. The safest policy might be to choose a system that is already in use overseas, but the designer should avoid that type of imitation.

G: Perhaps the simplest system is the best choice, but it is the designer's job to determine this. There are no rules for making the right decision. The choice has to be made on the basis of the specific technology, but specific technology means knowledge of the actual system, which comes down to imitation. Specific technology is not exactly the right term -- there's no word in English for this, and the closest is pertinent technology. I suppose that is because it is understood that you have to create the specific technology yourself. Many engineers in developing countries are familiar with the literature and draw on technology from overseas, so they know the specific technology very well, but I can't help but feel that this knowlege is an obstacle that prevents them from thinking up new systems and performing parameter design. Most people are tied to existing technology. The ideal person would be one who knew the specific technology and, precisely because he knew it, did not use it but tried his hand at a completely new system.

S: A new system might be a bridge that does not appear in the literature, or a completely new circuit. If the designer could derive the theoretical formulas for using this bridge or circuit to measure an unknown resistance y, he could perform parameter design as described in this chapter. In many cases, the problem is that the theoretical formulas do not exist. What should he do then?

G: For total system design, the subsystem output characteristics are the lower-level characteristics, and the total system output characteristics are the higher-level characteristics. The total system designer for a television set, for instance, wants to know what effect the picture tube, the power supply, the tuner, and the other subsystems will have on the sound and picture output characteristics. Questions such as what the variability and deterioration characteristics of current picture tubes are and how much changes in the supply voltage would cause problems are questions of tolerance design. The designer would have general answers to these questions, but that would not enable him to perform parameter design. For the functional characteristics of the picture tube, I would suggest that he should try reducing the voltage and power to half their present values, or doubling the stability with regard to variability or deterioration rate or the supply voltage to see how much these changes improve the quality.

S: But if you start making arbitrary demands like cut the power consumption in half or double the present stability, how will the subsystem designers be able to get the picture tube designed on schedule?

G: But these are only reference guidelines for quality evaluation; the designer is not being asked to actually develop a subsystem with half the power consumption and twice the stability. He is just being given guidelines for evaluating the quality improvement if such a subsystem could be developed.

S: I thought the guidelines for evaluating quality were the consumer tolerance $\pm \Delta_0$ and the loss A_0 caused by being outside the tolerance limits, or for a subsystem, the output characteristic specification m $\pm \Delta$ and the cost.

G: They are, but total system design, subsystem design, and so on are frequently performed in parallel because it is difficult, unless there is an analogous product already in existence, to estimate the loss A caused by not meeting the tolerances of the lower-level characteristics. The ideal is to develop a completely new system to provide the required functions at each design step. New systems are protected by patents.

S: But new systems involve a great deal of uncertainty.

G: The role of parameter and tolerance design is to eliminate that sort of technological uncertainty. The designer seeks the cheapest possible components and materials for the system he has conceived, then uses parameter design to improve the quality.

S: The first goal being to make it cheap.

G: That's right. Cost comes first, not quality. First get the cost cheap, then improve the quality by parameter design, then perform tolerance design after the optimum parameter levels have been set.

S: In the Wheastone bridge in this chapter, parameter design alone reduced the variance of the measurement error by a factor of more than 100, still using a third-grade galvanometer. I was surprised to see that much improvement -- more than with a first-grade galvanometer.

G: The designer's major effort should go into parameter design. It would be an extreme undertaking for even a large firm to run design studies of a large number of systems in parallel. Usually the design studies are limited to two or three systems at most. For the system or systems the designer selects, he should conduct the most thorough parameter design study possible given the time and UMC(Unit Manufacturing Cost) constraints. The materials, shapes, dimensions, characteristic values, and other values written on the drawings or in the specifications are all controllable factors in the parameter design study. What is wanted is not an effect curve showing how the output characteristics change in response to changes in the parameter values but information on the impact of various noise sources. It is unlikely that the designer will know the total interaction with all the error factors.

S: I can appreciate that. If he does not know the interaction with individual error factors, he can hardly know the interaction with all the error factors.

G: A more important point is that, in most designs, it is not clear what error factors will cause trouble. In 1980 I helped out in LSI research at Bell Laboratories. We wanted a 3.00 ± 0.25 μm window, but there was so much variability in the window size that we sometimes got no window at all. It was not clear what was causing all this variability. I told them there was no point in even thinking about the causes of window size variability; that the right strategy was to find a solution by changing the levels of the currently fixed parameters.

S: But there are innumerable fixed factors. Where would you start? You couldn't study all the parameters specified on the drawings.

G: That's a good question. When you're having trouble because there are too many controllable factors whose levels can be set freely, it's important to know where to start. I advised them to exclude the factors that could not be tested readily and list the ones that would be easy to study. If there were still too many, my advice was then to pick those easily testable parameters they thought were important. When you have theoretical formulas, of course, as in the example in this chapter, there are so few parameters that you should study them all. With the L_{54} orthogonal array you can design 25 parameters, and with the L_{108} array you can design 53. You can also use the L_{36} array three or four times instead of using the L_{108} array.

S: And I guess you can leave the controllable factors you don't study fixed, but what about the error factors? It would be strange to create error levels only for the control factors being studied. Can you discuss the layout problem a little more?

Layout of Error Factors

G: Error factors fall into three basic groups:

(1) External noise, or differences in the conditions under which the product is used: temperature, humidity, power supply voltage, and individual differences.

(2) Deterioration: the product is all right at first, but alters during storage or use, causing trouble due to variation or deterioration of its functions.

(3) Unit-to-unit variability: when one product functions properly but another does not, for instance.

Production and production engineering departments are concerned only with the factors in group (3), but the design department also has to deal with external noise (1) and deterioration (2).

S: Quality control both in the United States and at most Japanese firms was originally directed only toward unit-to-unit variability (3). Then, because of the importance of reliability, people started wanting to conduct life tests under various environmental conditions, and many firms built controlled-environment chambers to be able to change the environmental error factors under which functional tests are conducted.

G: But life tests and environmental tests are inefficient because they are time-consuming and costly. The design study method presented in this chapter did not consider either environmental error factors or deterioration factors. Presumably, if the environmental factors change, that will affect the measurement accuracy by affecting the element characteristics. Similarly, the deterioration of elements shows up as changes in the resistance values of resistors or the voltage of the power supply, which again are changes in internal constants. The three levels of the error factors can therefore represent variability in the characteristic values of the elements plus variability due to environmental changes and deterioration.

S: So in the design stage, the design study has to include all the error factors.

G: What I'm saying is that system design, parameter design, and tolerance design should aim for overall improvement in variability due to all error factors, not that all the error factors should be studied individually. When the S/N ratio can be computed from the outer orthogonal array in less than one second, as in theoretical design calculations, then as many error factors should be studied as possible. When experiments are necessary, however, only a small number of major error factors should be studied. There is, after all, a limit to the number of experiments that can be done.

S: I don't see how a small number of error factors can be enough when the environmental conditions change in many ways, or many system parameter values undergo deterioration.

G: That is a difficult objection to answer. It will be discussed in

Chapters 8 and 9, but in some cases, there can be experiments aimed at controlling error factors that do not include even one error factor. The assumption is that the level of a controllable factor that reduces the collective effect of a number of different error factors will reduce the effect of other error factors as well.

S: In other words, what will be resistant to the effects of one set of error factors will also be resistant to the effects of other error factors.

G: That's the argument, although it may be hard to accept. Deterioration of the resistance values of each resistor was studied as a separate error factor in this chapter, but if there are components that deteriorate together, their deterioration could be treated as a single error factor. The ± should be selected to increase output from engineering knowledge.

Level 1: initial R
Level 2: R ± 0.5%
Level 3: R ± 1.0%

If for all the components the three levels are the initial nominal value, ±0.5% the nominal value, and ±1.0% the nominal value, only one error factor is needed no matter how many components there are. This is called compound noise factor.

S: Environmental conditions and deterioration have to be considered in deciding the levels of the error factors, but the important point is to work with inexpensive components. Many designers try to start by improving quality, and not very many people take the approach of first getting the cost down then improving quality by parameter design.

G: As I said before, cost, not quality, comes first. The purpose of parameter design is to improve quality while staying at the minimum cost. Tolerance design should be performed only if the quality cannot be attained by parameter design alone.

S: What do you mean by "only if the quality cannot be attained?"

G: Tolerance design uses elements with small variability and deterioration, so it raises the cost whenever it is performed.

S: I know. Prototyping and testing with high-grade components and low-deterioration elements and materials in the key places is a mistake. But parameter design does not reduce the error variance to zero, so when the error loss is calculated after parameter design, the result is not zero. If σ^2 is the error variance after parameter design, the loss is:

$$L = k\sigma^2 \qquad\qquad ......(D\text{-}6.1)$$

Are you saying that tolerance design is necessary whenever this value is not zero?

G: A tolerance design study would be advisable, but that is not to say that a higher-grade element or material should be used. It only means that the tradeoff between cost and quality should be calculated. If the loss reduction attained by reducing the error variance by using slightly higher- priced second-grade elements instead of third-grade elements outweighs the cost increase, the second-grade elements should be used. If parameter design has been thoroughly carried out, however, the error variance σ^2 in equation D-6.1 should be quite small. If it is small already, reducing it to one-fourth of its value does not have much economic effect.

S: In the example in the text, without parameter design, it was advantageous to use a first-grade galvanometer, but after parameter design, the third-grade galvanometer was better.

G: That shows why it is extremely important to perform parameter design before tolerance design. A designer trying to make a good prototype will use expensive elements from the outset where he thinks it important. This may be all right for the initial prototype, but since parameter design should come first, a study should be made afterwards to select the optimum combination of parameter levels that will give stability and reliability even if the parameter characteristics vary, then the stability and reliability should be studied and tolerance design performed.

S: The old practice was to base the functional design on an initial prototype and then to study reliability and stability and correct any problems by requesting better components or elements, but I guess this practice is wrong.

G: That is the NASA method. It is one way to design a very reliable product, but the product will be too expensive to compete in free markets.

S: Design standards should also be reexamined from that point of view. Design standards and safety factors always seem to stress quality and drive the price of the product up.

G: If there is not enough time to do a design study, the designer should draw on the literature or design standards, but it will make for a less- competitive product. The preferable way is to start by making the product cheap, then improve its quality.

S: We've seen that there is a need to change the way design studies are done. if your costs are too high, you can't compete. The right way is to get the cost down, improve quality as much as possible by parameter design, then perform tolerance design if necessary at the end.

G: You might have perfect quality, with zero defectives produced and no problems in the field, but if the price were too high, the competition would still beat you. Price competition is primary. Low cost should be a precondition in raising quality levels.

S: It seems hard for people in other countries to understand this. That may be what leads people to think that Japanese factories are manned by workaholic workers driven to achieve quality even at additional cost. Is there an easy way to explain parameter design?

G: I don't think so, but the following example might convey some of the idea. Suppose one of the causes of variability is a temperature x in the production process. It's effect on the higher-level parameter y is not actually linear, but let's suppose that it is, and that y is a dimension. Under conditions that permit approximation by the linear regression equation

$$y = m + \beta (x - x) + e \qquad \qquad ......(D\text{-}6.2)$$

the variance σ_y^2 of y and the variance σ_x^2 of x are related by the formula

$$\sigma_y^2 = \beta^2 \sigma_x^2 + \sigma_e^2 \qquad \qquad ......(D\text{-}6.3)$$

where σ_e^2 is the variance of the error due to other factors. The idea of tolerance design is to reduce σ_y^2 by reducing σ_x^2. This method controls the source of the variation, although at an increased cost. Parameter design reduces the value of β^2. This method searches among the innumerable parameters that interact with x to find an inexpensive way to reduce β^2.

S: That explanation may help convince people who only know traditional statistical methods.

G: In the tile experiment conducted by Ina Seito in 1953, a way was found to reduce the dimensional variability due to temperature differences in the tunnel kiln to a fraction of its former value by adding a certain type of lime. In actual engineering studies, not only is the cause unknown, but it is frequently not even known in the design stage what kinds of internal and external noise are generated in the production process and during use.

S: So the designer has to find solutions to quality and cost problems caused by a large number of factors including those he knows nothing about.

G: That's right. This is precisely where the techniques of experimental design, orthogonal arrays, the S/N ratio, and the accumulation method prove useful. The real purpose of orthogonal arrays is to lay out the main effects and find how to make use of their interactions with unknown causes. Engineering is a completely different world from science with its obsession with mathematical models. This is the reason why many scientists are temperamentally unsuited for engineering development.

7 | *EXPERIMENTAL DESIGN FOR SMALLER-IS-BETTER CHARACTERISTICS*

Smaller-is-Better Characteristics and the Loss Function

A smaller-is-better characteristic is one that does not take on negative values and has a target of zero. The loss function L of a smaller-is-better characteristic is:

$$L = \frac{A_0}{\Delta_0^{\,2}} \, \sigma^2 \qquad\qquad ......(7.1)$$

where Δ_0 = consumer upper tolerance limit
A_0 = loss to society when the upper tolerance limit is exceeded
σ^2 = mean square error (variance)

Instead of Δ_0 and A_0, the following Δ and A can be used to perform the calculation:

Δ = shipping specifications
A = in-factory loss when the shipping specifications are not satisfied

Since Δ is given by:

$$\Delta = \sqrt{\frac{A}{A_0}} \times \Delta_0 \qquad\qquad ......(7.2)$$

we have the following equation.

$$\frac{A}{\Delta^2} = \frac{A_0}{\Delta_0{}^2} \qquad\qquad ......(7.3)$$

The purpose of this chapter is to explain how to perform experiments to improve a smaller-is-better characteristic and how to compare the data.

S/N Ratio of Smaller-is-Better Characteristic

The following is a problem from the second month of a correspondence course in experimental design given in China.

Problem: The data in the table below show measurements of wear on the spring side of an air pump slider after a durability test. Eight sliders were made under each of two conditions of surface finishing:

A_1 Shot-blast
A_2 Sandpaper

The measurements are in millimeters. Wear of more than 0.2 mm leads to trouble in the field and causes a loss of 600 yuan($300).

Table 7.1   Wear Data

									Total
A_1	0.09	0.13	0.05	0.04	0.08	0.08	0.07	0.05	0.59
A_2	0.03	0.05	0.05	0.04	0.04	0.02	0.01	0.04	0.28

For a smaller-is-better characteristic, if the data are y_1 through y_n, the variance σ^2 can be found as:

$$\sigma^2 = \frac{1}{n}(y_1{}^2 + y_2{}^2 + \cdots + y_n{}^2) \qquad\qquad ......(7.4)$$

By substituting the above value into the equation below, we can compare the quality obtained from A_1 and A_2.

$$L = \frac{A_0}{\Delta_0{}^2}\sigma^2 \qquad\qquad ......(7.5)$$

For A_1, the loss is calculated as:

$$\sigma^2 = \frac{1}{8}[0.09^2 + 0.13^2 + \cdots + 0.05^2] = 0.00616 \quad (mm^2)$$
$$......(7.6)$$

$$L = \frac{600}{0.2^2} \times 0.00616 = 92.4 \text{ yuan}$$

......(7.7)

For A_2:

$$\sigma^2 = \frac{1}{8}[0.03^2 + 0.05^2 + \cdots + 0.04^2] = 0.00140 \quad (\text{mm}^2)$$

....(7.8)

$$L = \frac{600}{0.2^2} \times 0.00140 = 21 \text{ yuan}$$

......(7.9)

It follows that A_2 is superior in quality to A_1 by:

$$92.4 - 21 = 71.4 \text{ yuan}$$

......(7.10)

If the cost of A_1 and A_2 is different, the cost difference can be subtracted from the quality difference before making the comparison.

The usual custom is to perform the following type of analysis of variance on wear data for A_1 and A_2.

$$S_m = \frac{0.87^2}{16} = 0.0473 \quad (f = 1)$$

......(7.11)

$$S_A = \frac{(0.59 - 0.28)^2}{16} = 0.0060 \quad (f = 1)$$

......(7.12)

$$S_T = 0.09^2 + 0.13^2 + \dots + 0.04^2 = 0.0605 \quad (f = 16)$$ ......(7.13)

$$S_e = S_T - S_m = 0.0605 - 0.0473 - 0.060 = 0.0072 \quad (f = 14)$$

......(7.14)

This gives the analysis-of-variance table shown as Table 7.2.

Table 7.2   Analysis of Variance

Source	f	S	V	$E(V)$	ρ (%)
m	1	0.0473	0.0473	$\sigma^2 + 16m^2$	77.3
A	1	0.0060	0.0060	$\sigma^2 + 8\sigma_A^2$	9.1
e	14	0.0072	0.00051	σ^2	13.6
T	16	0.0605			100.0

Taking what the analysis of variance table indicates to be the significant factor, we can calculate the average wear for A_1 and A_2 as follows.

$$\overline{A}_2 = \frac{0.28}{8} \pm \sqrt{\frac{4.60 \times 0.00051}{8}} = 0.035 \pm 0.017 \qquad \dots\dots(7.15)$$

$$\overline{A}_1 = \frac{0.59}{8} \pm \sqrt{\frac{4.60 \times 0.00051}{8}} = 0.074 \pm 0.017 \qquad \dots\dots(7.16)$$

The average wear for A_2 is therefore 0.039 mm less than that for A_1.

$$0.074 - 0.035 = 0.039 \text{ mm} \qquad \dots\dots(7.17)$$

Quality and Data

Since the loss from wear is related to the mean value, the way in which the conclusion in equation 7.17 was reached is not completely incorrect. For a smaller-is-better characteristic, a small mean means a small variability (standard deviation). If a small mean indicates a small standard deviation, then there is considerable justification for simply comparing the means. This makes much more sense than the irrational idea, found in some American statistics textbooks, of comparing means under the assumption of equal variances. Even when the means are equal, however, the variability may differ. An analysis that includes both the mean value and the variability is preferable.

Since the effects of internal noise, external noise, and other error factors all appear in the variability, quality must be measured in a way that includes the size of error. For a smaller-is-better characteristic, if the design life is T years, the average of the variance σ^2 over a period of T years should be estimated for a variety of conditions of use.

The variance when there is wear of β mm per year is given by the following equation, as explained in Chapter 3.

$$\frac{1}{T}\int_0^T (\beta t)^2 dt = \frac{T^2}{3}\beta^2 \qquad \dots\dots(7.18)$$

The actual value per year of the wear β differs according to the conditions under which the product is used. If β_1 through β_n are the observed values of β, then their mean square is

$$\text{mean } \beta^2 = \frac{1}{n}(\beta_1^2 + \beta_2^2 + \cdots + \beta_n^2) \qquad \dots\dots(7.19)$$

and this multiplied by $T^2/3$ is the estimated variance σ^2.

$$\sigma^2 = \frac{T^2}{3} \times (\text{mean } \beta^2) \qquad\qquad ......(7.20)$$

T is the design life. The mean β^2 ought to include variations in β due to different conditions of use as well as variations due to unit-to-unit differences. Table 7.1 does not include data for different conditions. To see the variability due to conditions of use, environmental factors should be taken as the error factors and an L_8 array used to carry out wear tests under eight sets of conditions for a period of $T/\sqrt{3}$ years. With this data, Table 7.1 would be complete, and the comparison made in equations 7.15, 7.16, and 7.17 would give the correct result.

If reducing the average wear in the wear tests by half would also reduce the standard deviation of the wear due to different conditions of use by half, then comparing the means would be equivalent to comparing the square root of the variances, or the standard deviation of the wear, or the "standard wear" for short. This will be further discussed in Chapter 9.

If the common logarithms of the mean values of A_1 and A_2 (0.074 and 0.035) are multiplied by -20, decibel quantities can be calculated as follows:

$$-20 \log\overline{A}_1 = -20 \log 0.074 = +22.6 \text{ decibels} \qquad ......(7.21)$$

$$-20 \log\overline{A}_2 = -20 \log 0.035 = +29.1 \text{ decibels} \qquad ......(7.22)$$

We can then speak of the gain of A_2 being higher than the gain of A_1 by:

$$29.1 - 22.6 = 6.5 \text{ decibels} \qquad\qquad ......(7.23)$$

Alternatively, one could say that finishing method A_2 gives a standard wear 6.5 decibels less than A_1.

The mean and standard deviation are not always related linearly, however, so instead of using the mean value alone, it would clearly be better to find the variance and multiply its common logarithm by -10 to compare quality.

Decibel value of A_1:
$$-10 \log\sigma^2 = -10 \log 0.00616 = 22.1 \text{ decibels} \qquad ......(7.24)$$

Decibel value of A_2:
$$-10 \log\sigma^2 = -10 \log 0.00140 = 28.5 \text{ decibels} \qquad ......(7.25)$$

This indicates that finishing method A_2 has a higher gain than A_1 by:

$$28.5 - 22.1 = 6.4 \text{ decibels} \qquad \qquad ......(7.26)$$

This value is almost the same as that given by equation 7.23.

In actual experiments, it would cost too much to attain high validity by testing all the relevant error factors, so the experiments should be conducted with a more limited range of factors. If the gain calculated from the variance for this range of factors is nearly equal to the gain in actual use, the experiments will yield valid optimum conditions, apart from cost.

The goal in research and development work and product design work should be to find the optimum production conditions or the optimum design for a wide variety of conditions of use by conducting beaker-scale tests, small-scale tests, instead of conducting tests under standard conditions, and full-scale tests. If problems turn up after production begins or after the product is sold, it is too late to cope with them easily. The question is how to find the optimum conditions or optimum design in the laboratory or design room. The right policy is to vary the factors as much as possible in the laboratory and include the resulting product variation in comparing the variance.

On this criterion, the experiment described in Table 7.1 is bad because it does not vary the factors. But, the data vary somewhat even in a bad experiment, so the variance comparison method of evaluation given in equations 7.8 and 7.9, which takes this variation into account, is preferable to the mean comparison method of equations 7.15 and 7.16. In addition, the calculations for 7.8 and 7.9 are easier. The ordinary analysis as given in Table 7.2 has some validity, but comparing the variance as in equations 7.8 and 7.9 is a better method.

Design of an Experiment on Wear

As an example of experimental design for a smaller-is-better characteristic, we shall examine an experiment for wear in a pump. This is also an example of parameter design for quality improvement, and it is advisable in this type of experiment to test the product under identical conditions and use those wear data as a control. From the wear data for the product under the present conditions, the wear per year can be calculated, and from that the amount of wear for the design life.

The object is to find design measures to reduce wear on the slider pump. The design factors are listed below.

A. Materials   Two levels
B. Weight   Two levels

C. Surface roughness   Two levels
D. Clearance   Two levels
E. Side material   Two levels

These are assigned to an L_8 orthogonal array so that all the main effects and the $A \times B$ and $A \times C$ interactions can be found. The data are for wear (in microns) at eight points on the slider of the pump. For comparison purposes, the existing product was tested under the same conditions to obtain wear data.

Table 7.3   Layout and Data

	A	B	$A \times B$	C	$A \times C$	D	E	\multicolumn{8}{c}{Data (μm)}
	1	2	3	4	5	6	7	R_1	R_2	R_3	R_4	R_5	R_6	R_7	R_8
1	1	1	1	1	1	1	1	12	12	10	13	3	3	16	20
2	1	1	1	2	2	2	2	6	10	3	5	3	4	20	18
3	1	2	2	1	1	2	2	9	10	5	4	2	1	3	2
4	1	2	2	2	2	1	1	8	8	5	4	3	4	9	9
5	2	1	2	1	2	1	2	16	14	8	8	3	2	20	33
6	2	1	2	2	1	2	1	18	26	4	2	3	3	7	10
7	2	2	1	1	2	2	1	14	22	7	5	3	4	19	21
8	2	2	1	2	1	1	2	16	13	5	4	11	4	14	30
	\multicolumn{7}{l}{Present product(benchmark)}	17	22	7	12	10	8	18	25

(1)   Find the S/N ratio and perform analysis of variance.
(2)   Find the optimum conditions, and calculate the approximate gain as compared with the present product.
(3)   If wear at any location exceeds $\Delta_0 = 200\mu m$, trouble will occur in the field, causing a loss A_0 of ¥80,000. Find the improvement of the optimum design over the present design in monetary terms. The design life is 20 years.

Calculation of S/N Ratio and Analysis of Variance

Parameter design is the work of determining the optimum levels of the parameters in a design. For a smaller-is-better characteristic, the goal is to reduce both the mean value and the variation, so the S/N ratio η can be calculated using σ^2 in the following equation.

$$\eta = -10 \log \sigma^2 \qquad\qquad ......(7.27)$$

If wear at all locations on the slider in the pump is of approximately equal importance, the variance σ^2 can be found from the sum of the squares of the wear at each point. In experiment No. 1, the wear data for the eight points are:

12, 12, 10, 13, 3, 3, 16, 20

The variance σ^2 is therefore:

$$\sigma^2 = \frac{1}{8}[12^2 + 12^2 + 10^2 + \cdots + 20^2] = \frac{1231}{8} = 153.9 \qquad \text{......(7.28)}$$

The S/N ratio is accordingly:

$$\eta = -10 \log \sigma^2 = -10\log 153.9 = -21.9 \text{ decibels} \qquad \text{......(7.29)}$$

If the S/N ratio in decibels is similarly calculated for experiments 2 to 8, and a working mean of 20 decibels is added, the results are as in Table 7.4.

Table 7.4   Layout and S/N Ratios

Col No.	A 1	B 2	$A \times B$ 3	C 4	$A \times C$ 5	D 6	E 7	S/N ratio +20
1	1	1	1	1	1	1	1	−1.9
2	1	1	1	2	2	2	2	−0.6
3	1	2	2	1	1	2	2	5.2
4	1	2	2	2	2	1	1	3.5
5	2	1	2	1	2	1	2	−4.2
6	2	1	2	2	1	2	1	−1.7
7	2	2	1	1	2	2	1	−3.0
8	2	2	1	2	1	1	2	−3.3
				Present product				−4.1

The S/N ratio is a characteristic value that represents quality. With the S/N ratio as the objective characteristic, analysis of variance can be carried out as follows.

$$S_T = (-1.9)^2 + (-0.6)^2 + \cdots + (-3.3)^2 - \frac{(-0.6)^2}{8} = 79.18 \qquad \text{......(7.30)}$$
$$(f = 7)$$

$$S_A = \frac{(6.2 + 12.2)^2}{8} = 42.32 \qquad (f = 1) \qquad \text{......(7.31)}$$

$$S_B = \frac{(-8.4 - 2.4)^2}{8} = 14.58 \qquad (f = 1) \qquad \text{......(7.32)}$$

$$S_{A \times B} = \frac{(-8.8-2.8)^2}{8} = 16.82 \qquad (f=1) \qquad \text{......(7.33)}$$

$$S_C = \frac{(-3.9+2.1)^2}{8} = 0.40 \qquad (f=1) \qquad \text{......(7.34)}$$

$$S_{A \times C} = \frac{(-1.7+4.3)^2}{8} = 0.84 \qquad (f=1) \qquad \text{......(7.35)}$$

$$S_D = \frac{(-5.9+0.1)^2}{8} = 4.20 \qquad (f=1) \qquad \text{......(7.36)}$$

$$S_E = \frac{(-3.1+2.9)^2}{8} = 0.01 \qquad (f=1) \qquad \text{......(7.37)}$$

From these values, it is possible to derive the analysis-of-variance table shown below (Table 7.5). Accuracy is low due to the small number of degrees of freedom, and the small factors C, $A \times C$, and E have been pooled as an error variation e.

Table 7.5 Analysis of Variance

Source	f	S	V	F_0	ρ (%)
A	1	42.32	42.32	100.8	52.9
B	1	14.58	14.58	34.7	17.9
$A \times B$	1	16.82	16.82	40.0	20.7
C	1	0.40	0.40	—	—
$A \times C$	1	0.84	0.84	—	—
D	1	4.20	4.20	10.0	4.9
E	1	0.01	0.01		
(e)	(3)	(1.25)	0.42		3.6
T	7	79.18			100.0

Estimate of Significant Factors and Conclusion

The significant factors do not have to be determined by matching them with an F table; a variance ratio of 2 or above can be taken as a criterion of significance. Here A, B, $A \times B$, and D will be considered significant, and their effects estimated. Since $A \times B$ is significant, estimates are made for all the combinations of A and B.

$$\overline{A_1 B_1} = \frac{-2.5}{2}' = -1.25 \ \Big]$$

$$\overline{A_1 B_2} = \frac{8.7}{2} = 4.35$$

$$\overline{A_2 B_1} = \frac{-5.9}{2} = -2.95 \qquad \qquad \text{......(7.38)}$$

$$\overline{A_2 B_2} = \frac{-6.3}{2} = -3.15$$

$$\overline{D_1} = \frac{-5.9}{4} = -1.48$$

$$\overline{D_2} = \frac{-0.1}{4} = -0.02 \qquad \qquad \text{......(7.39)}$$

The optimum conditions including non-significant factors are $A_1B_2C_2D_2.E_2$ The decibel value μ for these conditions is found as follows, where T is the overall average (Non-significant factors are not used for the estimation to avoid an over estimating).

$$\hat{\mu} = \overline{T} + (\overline{A_1}\overline{B_2} - \overline{T}) + (\overline{D_2} - \overline{T}) = \overline{A_1}\overline{B_2} + \overline{D_2} - \overline{T}$$
$$= 4.35 - 0.02 - (-0.75) = 5.08 \qquad \qquad \text{......(7.40)}$$

In comparison with the decibel value of -4.1 of the present product, this is a gain of:

$$5.08 - (-4.1) = 9.18 \text{ decibels} \qquad \qquad \text{......(7.41)}$$

Actual Value $= 10^{0.918} = 8.28$

The corresponding actual value is 8.28, so in comparison with the present product, wear-caused trouble should be reduced by a factor of 8.28.

If under actual field conditions the variance of the annual wear of the present product is:

$$o^2 = 28.0 \, \mu m^2 \qquad \qquad \text{......(7.42)}$$

Then the loss due to wear for the design life of $T = 20$ years is:

$$L = \frac{A_0}{\Delta_0^2} \times \frac{T^2}{3} \times \sigma^2 = \frac{80000}{200^2} \times \frac{20^2}{3} \times 28.0 = ¥ 7,467 \qquad \text{......(7.43)}$$

The optimum design $A_1B_2D_2$ will reduce this loss by a factor of 8.28, to:

$$L = ¥ 7,467 \times \frac{1}{8.28} = ¥ 902 \qquad \qquad \text{......(7.44)}$$

For each unit, this is a quality improvement of:

$$7{,}469 - 902 = ¥6{,}567 \qquad\qquad ......(7.45)$$

If the annual production volume is 200,000 units, the quality improvement amounts to ¥ 1.3 billion each year.

If there are no data for the wear of the present product, it is necessary to estimate roughly the number of years the experiment represents. The estimate can be off by 20% to 30% without seriously affecting the optimum conditions, because the cost difference between different levels is usually quite small.

PROBLEM

To reduce the CO content of exhaust gas, seven factors, A, B, C, D, E, F, and G are studied. Two levels are set for each and they are assigned to an L_8 orthogonal array. CO data (in grams) are obtained for three driving modes as given below.

No.	A B C D E F G 1 2 3 4 5 6 7	R_1	R_2	R_3	η (Decibels)
1	1 1 1 1 1 1 1	1.04	1.20	1.54	− 2.1
2	1 1 1 2 2 2 2	1.42	1.76	2.10	− 5.0
3	1 2 2 1 1 2 2	1.01	1.23	1.52	− 2.1
4	1 2 2 2 2 1 1	1.50	1.87	2.25	− 5.6
5	2 1 2 1 2 1 2	1.28	1.34	2.05	− 4.1
6	2 1 2 2 1 2 1	1.14	1.26	1.88	− 3.3
7	2 2 1 1 2 2 1	1.33	1.42	2.10	− 4.4
8	2 2 1 2 1 1 2	1.33	1.52	2.13	− 4.6
				Total	− 31.2

This is a smaller-is-better characteristic, and can be analyzed by using -10 times the logarithm of the mean squared error as the S/N ratio η. For experiment No.1, for example, the S/N ratio η is:

$$\eta = -10\log\frac{1.04^2 + 1.20^2 + 1.54^2}{3} = -2.1 \text{ decibels}$$

(a) Perform analysis of variance using the S/N ratio η as the objective characteristic data.

(b)   Estimate the significant factors.

(c) Find the optimum conditions and estimate the process average.

Then from the S/N ratio for these conditions, find the mean squared error σ^2. If the CO specification limit for that mode is 1.5 g and the loss A when the specification is not satisfied is ¥ 8,000, what is the loss under the optimum conditions? That is, find σ^2 and L in the following formulas.

σ^2 = reciprocal of actual value corresponding to decibel value under optimum conditions

Loss function $L = \dfrac{8000}{1.5^2}\, \sigma^2$

DISCUSSION

The Problem of One-Sided Specifications

S (Student): The purpose of parameter design is to determine the shapes, sizes, characteristics, and other specification values of the materials, parts, and other system components. In the case of a one-sided specification, however, since there is no nominal value, it seems more logical to call the specification a tolerance. For example, we might be talking about surface roughness C and the range might have two levels, C_1 and C_2. It doesn't matter which is which, so let C_1 be rough and C_2 smooth. The rough value is the limit: anything not rougher than that passes. In other words, this is a one-sided specification: C_1 or less.

G (Genichi): It would be unusual for there to be an optimum surface roughness, a case in which you did not want the surface to be either rougher or smoother than an optimum value. Ordinarily the specification would be equal to or less than a certain degree of roughness. I guess the question is whether to call this tolerance design, or to call it parameter design in that it is setting the level of a parameter. I'm not obsessed about this point, but for smaller-is-better characteristics and larger-is-better characteristics, I would prefer to call it parameter design. When the characteristic varies by more than a certain value on both sides of the target value, both positive and negative tolerances have to be determined.

S: There are some tolerances, such as clearances, with values like:

$m-\ 0\ \mu$m

$m+\ 5\ \mu$m   ......(D-7.1)

For a smaller-is-better characteristic such as wear, could we say:
$$0 - 0 \ \mu\text{m}$$
$$0 + 10 \ \mu\text{m} \qquad\qquad ......(\text{D-7.2})$$

G: For a clearance, the characteristic value of the product could be negative. In that case you have a defective unit that cannot be assembled, but the distribution of the characteristic is two-sided; it has a negative side. That makes it different from a smaller-is-better characteristic.

S: I think surface roughness is one-sided, but I'll drop the argument here.

Loss Function and Variance

S: Wear in each of the eight locations causes a loss, but the loss is for the same unit, so shouldn't the control factors be compared by adding up those losses? Instead of the variance of the wear at the eight locations, shouldn't the sum of the squares itself be used to calculate the loss?

G: That's reasonable. It's the same as the idea that since a product has several characteristic values, loss functions should be found for each of them and added to get the total. In practice, however, we obtain the factor effects for each characteristic. If we consider wear at each of the eight locations to be a separate characteristic, separate S/N ratios will have to be determined to carry out the analysis, but that involves too much testing. If a design that reduces wear at one location reduces the wear at the other locations by the same amount, there is nothing wrong with comparing the variance of the wear at the eight locations.

S: Aren't you simply saying that it is more efficient to perform the analysis by grouping the data in the form of the variance than to analyze the data separately?

G: You could put it that way. If the tolerance Δ_0 at one of the locations or the corresponding loss A_0 differs from the other locations, it could be weighted accordingly in the variance.

S: If we use weighting, then with Δ_{0i} as the tolerance of R_i and A_{0i} as the loss if that tolerance is not satisfied, I guess we would have:

$$\sigma^2 = \frac{1}{8} \sum_{i=1}^{8} \frac{A_{0i}}{\Delta_{0\,i}^{2}} \, y_i^{2} \qquad\qquad ......(\text{D-7.3})$$

G: That's right, but since the variance is used for making a comparison, it makes no difference whether you divide by eight or not.

8 EXPERIMENTAL DESIGN FOR LARGER-IS-BETTER CHARACTERISTICS

Larger-is-Better Characteristics

A characteristic value that does not take on negative values and for which the most desirable value is infinity is called a larger-is-better characteristic. Although it is true that "larger is better" is true for characteristics such as thermal efficiency, fraction yield, and percent nondefective, the maximum value is 1 (or 100%), so these are not true larger-is-better characteristics. Amplification, power, strength, capacity, and other characteristics which have no target value except "as large as possible" are larger-is-better characteristics.

Loss Function of a Larger-is-Better Characteristic

Let y be a larger-is-better characteristic with loss function $L(y)$, where $(0 \leq y \leq \infty)$. In the neighborhood of $y = \infty$, $L(y)$ has the following Laurent expansion:

$$L(y) = L(\infty) + \frac{L'(\infty)}{1!} \times \frac{1}{y} + \frac{L''(\infty)}{2!} \times \frac{1}{y^2} + \cdots\cdots\cdots \quad \dots\dots(8.1)$$

At $y = \infty$, the loss is zero, and this is the minimum value of the loss, so it is reasonable to stipulate that:

$$L(\infty) = 0 \qquad \qquad \dots\dots(8.2)$$
$$L'(\infty) = 0 \qquad \qquad \dots\dots(8.3)$$

It follows that the first term in the expansion (8.1) of the loss function is:

$$L(y) \doteq k \times \frac{1}{y^2}$$

.....(8.4)

This can be used as an approximation of the loss function. If the loss $L(y_0)$ at any point $y = y_0$ is known, the coefficient k can be found as:

$$k = y_0^2 \times L(y_0)$$

....(8.5)

The point y_0 can be taken as the point Δ_0 at which trouble actually occurs in the field. If the loss at that point is A_0, we have:

$$k = \Delta_0^2 \times A_0$$

......(8.6)

If there is a higher-level characteristic with a specification limit of not less than Δ_0, and the loss when this specification is not satisfied is A_0, then the coefficient in the loss function for that characteristic can be found from the formula:

$$k = \Delta_0^2 A_0 / b^2$$

......(8.7)

where b is the coefficient of the effect on the higher-level characteristic of a unit change in the lower-level characteristic. The loss function is accordingly:

$$L = \frac{\Delta_0^2 A_0}{y^2} \quad \text{or} \quad \frac{\Delta_0^2 A_0}{y^2} \times \frac{1}{b^2}$$

......(8.8)

From this, the shipping specifications are given by the following formulas:

$$\left. \begin{array}{l} \Delta = \sqrt{\dfrac{A_0}{A}} \times \Delta_0 \quad \text{or more} \\[3ex] \text{or} \\[1ex] \Delta = \sqrt{\dfrac{A_0}{A}} \times \dfrac{\Delta_0}{b} \quad \text{or more} \end{array} \right\}$$

......(8.9)

Sample Calculation

To see how this works in actual practice, let us solve the following problem.

Problem: To determine the manufacturing specification (the specification on the drawings) for the joint strength of a pipe. If the joint strength is less than 12 kg, the pipe will fail during use, and the average value of the loss A_0 is about ¥ 200,000. The in-factory loss A when the specification limit is not satisfied is ¥ 3,000. Find the manufacturing specification.

Solution: In this problem, the tolerance limit in the field is:

$$\Delta = 12kg$$

and the loss from a failure in the field is:

$$A_0 = ¥ 200,000$$

Since the in-factory loss caused by not satisfying the specification limit is $A = ¥ 3,000$, the loss function L is:

$$L = \Delta_0^2 A_0 \frac{1}{y^2} = 12^2 \times 200000 \times \frac{1}{y^2} = \frac{28800000}{y^2} \qquad ......(8.10)$$

The in-factory loss A when the specification is not satisfied can be substituted for the left side to solve for y.

$$3000 = \frac{28800000}{y^2} \qquad ......(8.11)$$

Accordingly, the drawing specification Δ for y should be:

$$\Delta = \sqrt{\frac{28800000}{3000}} = 98 \qquad .....(8.12)$$

The strength should be at least 98 kg. Underlying this solution, of course, is the assumption that a pipe with a strength of 98 kg can be manufactured. The design task is to produce a design that can be manufactured at cost A. Different materials can be used, and the shape and thickness can be changed.

Experiment on Bonding Strength of Plastic Product, and S/N Ratio

The consumer tolerance Δ_0 for the bonding strength of a plastic product is 10 kg/cm^2, and the loss A_0 when the specification is not satisfied is ¥ 76,000.

To find the optimum bonding conditions, the factors listed in Table

8.1 were assigned to an L_{27} orthogonal array. Table 8.2 gives the measured bonding strength data.

Table 8.1 Factors and Levels

A. Etch time (min)	5	10	15
B. Etch temperature (ºC)	60	65	80
C. Etchant composition	C_1	C_2	C_3
D. Preprocessing	None	Solvent	Warm water
E. Accelerator	E_1	E_2	E_3
F. Catalyst	Present	Proposed$_1$	Proposed$_2$
G. Neutralizing method	G_1	G_2	G_3

Table 8.2 Layout and Data

Col No.	A	B	A×B	C	A×C			B×C	D	E	B×C	F	G	Data		Decibel value (η-20)
	1	2	3	4	5	6	7	8	9	10	11	12	13			
1	1	1	1	1	1	1	1	1	1	1	1	1	1	6	5	− 5.3
2	1	1	1	1	2	2	2	2	2	2	2	2	2	10	8	− 1.1
3	1	1	1	1	3	3	3	3	3	3	3	3	3	10	12	+ 0.7
4	1	2	2	2	1	1	1	2	2	2	3	3	3	3	10	− 7.8
5	1	2	2	2	2	2	2	3	3	3	1	1	1	18	18	5.1
6	1	2	2	2	3	3	3	1	1	1	2	2	2	23	18	6.0
7	1	3	3	3	1	1	1	3	3	3	2	2	2	9	13	0.4
8	1	3	3	3	2	2	2	1	1	1	3	3	3	33	30	9.9
9	1	3	3	3	3	3	3	2	2	2	1	1	1	29	29	9.2
10	2	1	2	3	1	2	3	1	2	3	1	2	3	6	8	− 3.4
11	2	1	2	3	2	3	1	2	3	1	2	3	1	7	11	− 1.6
12	2	1	2	3	3	1	2	3	1	2	3	1	2	23	24	7.4
13	2	2	3	1	1	2	3	2	3	1	3	1	2	1	1	−20.0
14	2	2	3	1	2	3	1	3	1	2	1	2	3	31	31	9.8
15	2	2	3	1	3	1	2	1	2	3	2	3	1	32	35	10.5
16	2	3	1	2	1	2	3	3	1	2	2	3	1	16	20	4.9
17	2	3	1	2	2	3	1	1	2	3	3	1	2	32	35	10.5
18	2	3	1	2	3	1	2	2	3	1	1	2	3	29	32	9.7
19	3	1	3	2	1	3	2	1	3	2	1	3	2	1	1	−20.0
20	3	1	3	2	2	1	3	2	1	3	2	1	3	37	34	11.0
21	3	1	3	2	3	2	1	3	2	1	3	2	1	33	28	9.6
22	3	2	1	3	1	3	2	2	1	3	3	2	1	13	16	3.1
23	3	2	1	3	2	1	3	3	2	1	1	3	2	37	35	11.1
24	3	2	1	3	3	2	1	1	3	2	2	1	3	31	33	10.1
25	3	3	2	1	1	3	2	3	2	1	2	1	3	28	28	8.9
26	3	3	2	1	2	1	3	1	3	2	3	2	1	35	38	11.2
27	3	3	2	1	3	2	1	2	1	3	1	3	2	36	35	11.0

All of the main effects and the interactions among A, B, and C were assigned to the L_{27} array. Two test pieces were manufactured for each set of conditions and a pull-off test was performed, yielding the data in

Table 8.2. For a larger-is- better characteristic, infinity is the ideal value. For the data y_1 and y_2 for the two pieces, the average of the reciprocal squares is:

$$\sigma^2 = \frac{1}{2}\left(\frac{1}{y_1{}^2} + \frac{1}{y_2{}^2}\right)$$

......(8.13)

From this, the S/N ratio η can be calculated:

$$\eta = -10 \log \sigma 2$$

......(8.14)

For example, from the data for experiment No. 1:

$$\sigma^2 = \left(\frac{1}{6^2} + \frac{1}{5^2}\right) \times \frac{1}{2} = 0.0339$$

......(8.15)

$$\eta = -10 \log \sigma 2 = -10 \log 0.0339 = 14.7 \text{ decibels}$$

......(8.16)

Since a working mean of 20 decibels was subtracted, the value given in Table 8.2 is -5.3 decibels.

Analysis of Variance

The analysis is performed using the decibel values. Since the layout was arranged to find the interactions of A, B, and C, besides the totals for A, B, C, D, E, F, and G, the totals of the two-way arrays for AB, AC, and BC are also found. These two-way arrays are given in Table 8.3.

Table 8.3   Supplementary Tables

Two-way arrays

$A\,B$	Total	$A\,C$	Total	$B\,C$	Total
1 1	$-\ 5.7$	1 1	-12.7	1 1	-28.7
1 2	3.3	1 2	13.9	1 2	8.3
1 3	19.5	1 3	15.9	1 3	17.7
2 1	2.4	2 1	-18.5	2 1	-24.7
2 2	0.3	2 2	18.7	2 2	26.0
2 3	25.1	2 3	27.6	2 3	26.6
3 1	0.6	3 1	$-\ 8.0$	3 1	14.2
3 2	24.3	3 2	33.3	3 2	31.6
3 3	31.1	3 3	30.7	3 3	29.9

Main effects

	A	B	C	D	E	F	G
1	17.1	− 2.7	− 39.2	57.8	28.3	36.9	46.7
2	27.8	27.9	65.9	47.5	23.7	45.3	5.3
3	56.0	75.7	74.2	− 4.4	48.9	18.7	48.9
Total	100.9	100.9	100.9	100.9	100.9	100.9	100.9

From Table 8.3, the variations can be found as:

$$S_A = \frac{17.1^2 + 27.8^2 + 56.0^2}{9} - \frac{100.9^2}{27} = 89.74 \qquad (f = 2) \qquad \text{......(8.17)}$$

$$S_B = \frac{(-2.7)^2 + 27.9^2 + 75.7^2}{9} - \frac{100.9^2}{27} = 346.95 \; (f = 2) \qquad \text{......(8.18)}$$

$$S_{AB} = \frac{(-5.7)^2 + 3.3^2 \cdots + 31.1^2}{3} - \frac{100.9^2}{27} = 495.45 \; (f = 8) \qquad \text{......(8.19)}$$

$$S_{A \times B} = S_{AB} - S_A - S_B = 495.45 - 89.74 - 346.95 - 58.76 \; (f=4) \qquad \text{......(8.20)}$$

Similarly:

$$S_C = 887.94 \qquad (f=2) \qquad \text{......(8.21)}$$

$$S_{A \times C} = 37.38 \qquad (f=2) \qquad \text{......(8.22)}$$

$$S_{B \times C} = 152.62 \qquad (f=2) \qquad \text{......(8.23)}$$

$$S_D = 246.98 \qquad (f=2) \qquad \text{......(8.24)}$$

$$S_E = 40.02 \qquad (f=2) \qquad \text{......(8.25)}$$

$$S_F = 41.09 \qquad (f=2) \qquad \text{......(8.26)}$$

$$S_G = 134.07 \qquad (f=2) \qquad \text{......(8.27)}$$

From these values the analysis-of-variance table given as Table 8.4 can be constructed:

Estimation of Factorial Effects

The analysis of variance in Table 8.4 enables us to estimate the major factors. The estimates can be read from Tables 8.3 and 8.5. A working mean of 20 decibels has been added in Table 8.5.

Table 8.4   Analysis of Variance

Source	f	S	V		F_0	ρ (%)
A	2	89.74	44.87		3.03	2.96
B	2	346.95	173.48		11.75	15.59
$A \times B$	4	58.76	14.69	○	—	
C	2	887.94	443.97		30.06	42.17
$A \times C$	4	37.38	9.34	○	—	
$B \times C$	4	152.62	38.16		2.58	4.60
D	2	246.98	123.49		8.36	10.68
E	2	40.02	20.01	○	—	
F	2	41.09	20.54	○	—	
G	2	134.07	67.04		4.54	5.14
e	0	0.00				
(e)	(1 2)	(177.25)	(14.77)			18.86
T	26	2035.55				100.00

Note : Circles indicate pooling.

Table 8.5   Estimates of Significant Factors

$\overline{A}_1$	$= 21.90 \pm 2.79$		$B_3 C_1 = 24.73 \pm 4.84$	
$\overline{A}_2$	$= 23.09 \pm 2.79$		$B_3 C_2 = 30.53 \pm 4.84$	
$\overline{A}_3$	$= 26.22 \pm 2.79$		$B_3 C_3 = 29.97 \pm 4.84$	
$B_1 C_1 = 10.43 \pm 4.84$			$\overline{D}_1$	$= 26.42 \pm 2.79$
$B_1 C_2 = 22.77 \pm 4.84$			$\overline{D}_2$	$= 25.28 \pm 2.79$
$B_1 C_3 = 25.90 \pm 4.84$			$\overline{D}_3$	$= 19.51 \pm 2.79$
$B_2 C_1 = 11.77 \pm 4.84$			$\overline{G}_1$	$= 25.19 \pm 2.79$
$B_2 C_2 = 28.67 \pm 4.84$			$\overline{G}_2$	$= 20.59 \pm 2.79$
$B_2 C_3 = 28.87 \pm 4.84$			$\overline{G}_3$	$= 25.43 \pm 2.79$

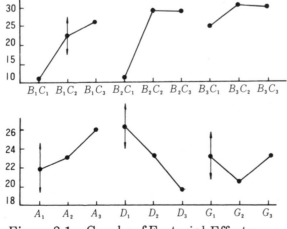

Figure 8.1   Graphs of Factorial Effects

Although the confidence limits are not important, it may be useful to know how they are derived. For the main effects, in the F table for 1 and 12 degrees of freedom, the 5% value is 4.75.

$$\pm \sqrt{\frac{F \times V_e}{n_e}} = \pm \sqrt{\frac{4.75 \times 14.77}{9}} = \pm 2.79 \qquad \text{......(8.28)}$$

For the B,C combination,

$$\pm \sqrt{\frac{4.75 \times 14.77}{3}} = \pm 4.84 \qquad \text{......(8.29)}$$

Determination of Optimum Conditions and Process Average

From Figare 8.1 and Tables 8.5, the optimum levels are $A_3B_3C_2D_1E_3F_2G_3$. Under these conditions the process average $\hat{\mu}$ is:

$$
\begin{aligned}
\hat{\mu} &= \overline{A}_3 + \overline{B}_3\overline{C}_2 + \overline{D}_1 + \overline{G}_3 - 3\overline{T} \\
&= 26.22 + 30.53 + 26.42 + 25.43 - 3 \times 23.74 \qquad \text{......(8.30)} \\
&= 37.38
\end{aligned}
$$

This value indicates that the mean square of the reciprocal of the bonding strength of the plastic product is 37.38 decibels. If the true mean square of the reciprocal of the bonding strength is σ^2:

$$
\begin{aligned}
-10 \log\sigma^2 &= 37.38 \\
\sigma^2 &= 0.000183 \qquad \text{......(8.31)}
\end{aligned}
$$

Substituting $A_0 = ¥76,000$ and $\Delta_0 = 10$, the values given in the problem, into the formula for the loss function gives:

$$L = A_0\Delta_0^2 \times \frac{1}{y_2} = k\sigma^2 \qquad \text{......(8.32)}$$

$$L = 76000 \times 10^2 \times \sigma^2 = 76000 \times 0.000183 = ¥1,391 \qquad \text{......(8.33)}$$

The initial set of conditions considered by the experimenters was $A_2B_2C_2D_2E_2F_1G_2$. We shall find the monetary improvement obtained by using the optimum conditions.

For $A_2B_2C_2D_2E_2F_1G_2$, the process average is:

$$
\begin{aligned}
\hat{\mu} &= \overline{A}_2 + \overline{B}_2\overline{C}_2 + \overline{D}_2 + \overline{G}_2 - 3\overline{T} \\
&= 23.09 + 28.67 + 25.28 + 20.59 - 3 \times 23.74 \\
&= 26.41 \qquad \text{......(8.34)}
\end{aligned}
$$

Accordingly,

$$\sigma^2 = 10^{-2.641} = 0.00229 \qquad \text{......(8.35)}$$

$$L = 7600000 \times 0.00229 = ¥\,17,404 \qquad \text{......(8.36)}$$

From this, the monetary improvement is:

$$17,404 - 1,391 = ¥\,16,013$$

If the annual production volume is 200,000 units, the improvement amounts to ¥ 3.2 billion annually.

PROBLEM

The specification for the bonding strength of a certain maker of epoxy resin is 3.00 tons or more. The rework cost A when this specification is not satisfied is ¥ 30,000. In an effort to improve the bonding strength, three levels were set for each of the following factors:

A = composition of bond
B = bonding method
C = surface treatment
D = worker

These factors were assigned to an L_9 orthogonal array, and two pieces were joined in each experiment. Measured data for bonding strength in tons are given below.

No.	A	B	C	D	Data		No.	A	B	C	D	Data	
1	1	1	1	1	6.80	2.27	6	2	3	1	2	2.93	2.72
2	1	2	2	2	2.49	3.43	7	3	1	3	2	1.70	2.12
3	1	3	3	3	2.17	1.57	8	3	2	1	3	4.24	1.91
4	2	1	2	3	1.79	1.33	9	3	3	2	1	1.50	4.05
5	2	2	3	1	1.98	2.57							

(1) Find the logarithm of the mean square of the reciprocals, multiply it by -10 to obtain decibel data, and perform analysis of variance.
(2) Find the significant factors, determine the optimum levels, and estimate the process average.

(3)   If the present levels of all the factors are level 2, what is the improvement in decibels for the optimum levels? Find the gain.

DISCUSSION

S/N Ratio and Loss Function

S (Student):Instead of calculating $\eta = m^2/\sigma^2$ as the *S/N* ratio, we are using the mean square of the reciprocal directly, which is the variance for the loss function. What is the problem here?

G (Genichi):The problem, for both the designer and the production engineer, is how much trouble will occur during the design life of a product under a variety of conditions of use. Suppose, for example, that there are two alternative materials or two designs, A_1 and A_2. If we make 100 of each and use them under varying conditions for an assumed design life of 20 years, we can add up the total amount of trouble from A_1 and A_2 and compare.

S: A_1 may be better under one set of conditions and A_2 better under another set of conditions, but the question is which is better on the average.

G: Right. There may be interactions with various of the conditions, but this information is meaningless. The question is which is better on the average. This is what we are calling the main effect.

S: Comparing products used for 20 years under varying conditions strikes me as a very inefficient way to do an experiment. In the example in the text, the test was not conducted under varying conditions, and it did not take degradation over 20 years into account. Wouldn't it be a bit difficult for the development personnel to know about those things?

G: That's the beauty of the orthogonal array. Deterioration and changes in the environmental conditions show up as changes in the characteristic values of the product. To develop a product that is resistant to environmental and deterioration effects, in theory, you should find the error variance which measures the size of the variability with respect to those conditions and calculate the S/N ratio, but the example in the text ignores error factors.

The reason is that the role of the error factors is taken over by the other controllable factors. A change in the level of a controllable factor will also cause the characteristic values of the product to change. When A_1, A_2, and A_3 are compared, the other factors B, C, D, and so on are also changing.

S: In other words, measures taken to deal with variation caused by changes in the controllable factors will also take care of variation caused by error factors.

G: That's right. For example, suppose you had the following strength data for A_1, A_2, and A_3 which are assigned to the L_{27} array. I've fixed this so that there are no repeated data.

Table D-8.1   Data

Level	Data									Total
A_1	24	10	32	5	38	15	9	40	7	180
A_2	17	24	15	27	20	16	26	18	17	180
A_3	20	20	20	20	20	20	20	20	20	180

S: If A is assigned to the first column, the first nine data values are for A_1, the next nine are for A_2, and the third nine are for A_3. In the data for each of A_1, A_2, and A_3, there are additional factors B, C, D, and so on that are changing.

G: The means are all the same. See if you can find the S/N ratios for A_1, A_2, and A_3.

S: That means finding the average of -20 times logarithm, which works out as:

$$\overline{A}_1 = \frac{20}{9}(\log 24 + \log 10 + \cdots + \log 7) = 23.87 \quad \text{decibels} \qquad ......(D\text{-}8.1)$$

$$\overline{A}_2 = \frac{20}{9}(\log 17 + \log 24 + \cdots + \log 17) = 25.83 \qquad ......(D\text{-}8.2)$$

$$\overline{A}_3 = \frac{20}{9}(\log 20 + \log 20 + \cdots + \log 20) = 26.02 \qquad ......(D\text{-}8.3)$$

These values can be summarized as shown in Table D-8.2.

Table D-8.2   Comparison of Levels by S/N Ratio

Level	S/N ratio	Converted mean
A_1	23.87	15.61
A_2	25.83	19.57
A_3	26.02	20.00

For A_1, the value converted to the mean (m) indicates what the mean value would be if all the data values were the same. It is calculated as:

$$20(\log m) = 23.87 \qquad\qquad ......(D\text{-}8.4)$$

$$m = 15.61 \qquad\qquad ......(D\text{-}8.5)$$

Since there are some small values like 5 and 7 in the data for A_1, I'd expect the value converted to the mean to be worse.

G: Find the variance of the loss function from Table D-8.1.
Calculate σ^2 separately for each of A_1, A_2, and A_3.

$$\sigma^2 = \frac{1}{9}\left(\frac{1}{y_1^2} + \frac{1}{y_2^2} + \cdots + \frac{1}{y_9^2} \right) \qquad\qquad ......(D\text{-}8.6)$$

S: That works out as:
For A_1,

$$\sigma^2 = \frac{1}{9}\left(\frac{1}{24^2} + \frac{1}{10^2} + \cdots + \frac{1}{7^2} \right) = 0.01014 \qquad\qquad ......(D\text{-}8.7)$$

For A_2

$$\sigma^2 = \frac{1}{9}\left(\frac{1}{17^2} + \frac{1}{24^2} + \cdots + \frac{1}{17^2} \right) = 0.00283 \qquad\qquad ......(D\text{-}8.8)$$

For A_3,

$$\sigma^2 = \frac{1}{9}\left(\frac{1}{20^2} + \frac{1}{20^2} + \cdots + \frac{1}{20^2} \right) = 0.00250 \qquad\qquad .....(D\text{-}8.9)$$

This comparison gives us Table D-8.3.

Table D-8.3   Comparison of Levels by Loss Function

Level	Variance	Decibel	Gain	Converted mean
A_1	0.01014	19.94	Base	9.93
A_2	0.00283	25.48	+5.54	18.80
A_3	0.00250	26.02	+6.08	20.00

The value converted to the mean comes from the square root of the reciprocal of the variance, indicating what the mean value would be if there were no variability.

If we compare Tables D-8.2 and D-8.3, Table D-8.3 shows larger differences in both the decibel value of the S/N ratio and the value converted to the mean. Which is more important?

G: We're not considering actual error factors (environmental factors, degradation, unit-to-unit differences). We're letting the other control variables substitute for the error factors. So we can't predict the actual loss function accurately.

S: Which means that it is not clear whether it is better to use Table D-8.2 or Table D-8.3.

G: True. It would help to select a judicious number of error factors, assign them to an outer array, and calculate the S/N ratio from the outer array data, but that would be an expensive and inefficient experiment. Since the data in Table D-8.2 do not include error factors, they probably underestimate the variability in the S/N comparison. I would recommend using the method of analysis in the text on the data for the standard conditions and one or two worst-case conditions. Table D-8.3 gives us variability due to controllable factors. If the difference between levels of the controllable factors is large enough, the variability due to controllable factors will necessarily be larger than the variability due to error factors.

S: But the difference from the error factors might be bigger. How can you be sure that the controllable factor effect is larger than the error factor effect?

G: The assumption is that the difference between the controllable factor levels is very large. The problem with the method in Table D-8.3 is that the error variance is confounded by the interaction of the main effects. When the method of Table D-8.3 is used, the orthogonal array should be L_{18}, L_{36}, L_{12}, L_{20}, something of that order. The point I want to make is that in finding the optimum conditions, the comparison should include the size of the variation, not just the mean values.

S: There seem to be problems with the calculations in both Table D-8.2 and Table D-8.3.

G: Yes, in this case. But they are clearly better than just comparing the means.

S: How about calculating the following two characteristics?

Sensitivity $\qquad \eta' = 10 \log Sm$ $\qquad\qquad$ ......(D-8.10)

S/N ratio $\qquad \eta = \dfrac{(S_m - V_e)/9}{V_e}$ $\qquad\qquad$ ......(D-8.11)

where

$S_m = \dfrac{(y_1 + y_2 + \cdots + y_9)^2}{9}$ $\qquad\qquad$ ......(D-8.12)

$$V_e = \frac{1}{8} \Sigma (y_i - \bar{y})^2$$

......(D-8.13)

G: You could analyze the mean values instead of η'. The problem is the difference in importance between the effect on η and the effect on η'. Besides determining the optimum conditions from η, which gives a measure of stability, the problem is what to do about the factors that are only significant for η' or the mean value. I think it would be adequate to consider at least one major error factor, or to use the analysis method in the text on the data for the standard conditions and one or two worst-case conditions.

S: But it would be better to select several major error factors, assign them to a L_9 array, find their S/N ratios as below, and use these S/N ratios in the analysis.

$$\eta = -10 \log \left[\left(\frac{1}{y_1^2} + \frac{1}{y_2^2} + \cdots + \frac{1}{y_9^2} \right) \times \frac{1}{9} \right]$$

......(D-8.14)

G: If cost were no concern, that would be ideal. Another possibility is the accumulation method of studying the distribution of the data, but in the interests of experimental efficiency, it might be better to use an L_{18} or L_{36} array with the analysis method of Table D-8.3, particularly in the case of destructive testing such as for strength characteristics.

S: It sounds like what you're saying is if the experiments are not too difficult, use the method of Section 8.2 in the text; if the data are for standard and worst-case conditions, use the method of Table D-8.2; and in other cases lay out an L_{18} or L_{36} array and use Table D-8.3.

G: That pretty well sums it up. Of course, you could also use the method in Chapter 9, too.

9

BY-PASSING THE S/N RATIO: SPRING EXPERIMENT

Life Test

When there is sufficient information on wear and deterioration of parts and materials, the design procedures described so far can be used by setting appropriate levels of wear and deterioration for the projected life of the product. When there is not sufficient information, it is necessary to use the *S/N* ratio method or conduct life tests. Life tests with respect to the final design can be conducted in parallel with production and sale, but improvement in life characteristics requires life testing during the design and prototype stages.

Since life tests cannot be repeated all that often, it is best to cover as many factors as possible in one series of experiments. To assure good stability and adequate life, external noise factors should be studied, but if this is not practical, the experiment should use a large inner array with as many controllable factors assigned as possible, as in the example in this chapter.

In a life test, at most two or three test pieces should be made for each experiment in the array, and the test should be continued until about half of the specimens have failed. The test period (including accelerated life tests) should be divided into about 10 levels, ω_1 through ω_k. At each time level, the data should be 1 if the specimen is still functioning and 0 if it is not. The minute analysis can be used.

Clutch Spring Life Test:Analysis of Variance

The following factors were studied in connection with the

durability of a clutch spring:

A. Shape, 3 levels
B. Hole ratio, 2 levels
C. Coining, 2 levels
D. Stress σ_t, 90 65 40
E. Stress σ_c, 200 170 140
F. Shot peening, 3 levels
G. Outer perimeter planing, 3 levels

These factors were assigned to an L_{27} array to find all the main effects, and the $D \times E$ and $D \times F$ interactions. The two-level factors B and C were combined into one three-level factor. Three springs were made for each of the 27 conditions and cycled 1,100,000 times to test durability. The levels in the data were:

Table 9.1 Layout and Data

Col. No.	D	E	D×E E	D×E E	A	B̂/C	e	F	D×F F	D×F F	G	e	e	ω_1	ω_2	ω_3	ω_4	ω_5	ω_6	ω_7	ω_8	ω_9	ω_{10}	ω_{11}	Total
	1	2	3	4	5	6	7	8	9	10	11	12	13												
1	1	1	1	1	1	1	1	1	1	1	1	1	1	3	0	0	0	0	0	0	0	0	0	0	3
2	1	1	1	1	2	2	2	2	2	2	2	2	2	3	3	3	3	2	1	1	1	1	1	1	20
3	1	1	1	1	3	3	3	3	3	3	3	3	3	3	3	1	1	1	1	1	1	1	1	1	15
4	1	2	2	2	1	1	1	2	2	2	3	3	3	3	3	2	0	0	0	0	0	0	0	0	8
5	1	2	2	2	2	2	2	3	3	3	1	1	1	3	3	3	3	3	2	2	2	2	2	2	27
6	1	2	2	2	3	3	3	1	1	1	2	2	2	3	0	0	0	0	0	0	0	0	0	0	3
7	1	3	3	3	1	1	1	3	3	3	2	2	2	3	1	1	0	0	0	0	0	0	0	0	5
8	1	3	3	3	2	2	2	1	1	1	3	3	3	3	1	0	0	0	0	0	0	0	0	0	4
9	1	3	3	3	3	3	3	2	2	2	1	1	1	3	3	3	1	0	0	0	0	0	0	0	10
10	2	1	2	3	1	2	3	1	2	3	1	2	3	3	1	0	0	0	0	0	0	0	0	0	4
11	2	1	2	3	2	3	1	2	3	1	2	3	1	3	3	3	3	3	3	3	3	3	3	3	33
12	2	1	2	3	3	1	2	3	1	2	3	1	2	3	3	3	3	3	3	2	2	2	2	2	28
13	2	2	3	1	1	2	3	2	3	1	3	1	2	3	3	3	3	3	3	3	3	3	3	3	33
14	2	2	3	1	2	3	1	3	1	2	1	2	3	3	3	0	0	0	0	0	0	0	0	0	6
15	2	2	3	1	3	1	2	1	2	3	2	3	1	3	2	0	0	0	0	0	0	0	0	0	5
16	2	3	1	2	1	2	3	3	1	2	2	3	1	3	3	2	1	0	0	0	0	0	0	0	9
17	2	3	1	2	2	3	1	1	2	3	3	1	2	3	3	0	0	0	0	0	0	0	0	0	6
18	2	3	1	2	3	1	2	2	3	1	1	2	3	3	3	3	3	3	3	3	3	3	3	3	33
19	3	1	3	2	1	3	2	1	3	2	1	3	2	3	3	3	2	0	0	0	0	0	0	0	11
20	3	1	3	2	2	1	3	2	1	3	2	1	3	3	3	3	3	3	3	3	3	3	3	3	33
21	3	1	3	2	3	2	1	3	2	1	3	2	1	3	3	3	3	3	3	3	3	3	3	3	33
22	3	2	1	3	1	3	2	2	1	3	3	2	1	3	3	3	3	3	3	3	3	3	3	3	33
23	3	2	1	3	2	1	3	3	2	1	1	3	2	3	3	3	3	3	3	3	3	3	3	3	33
24	3	2	1	3	3	2	1	1	3	2	2	1	3	3	3	3	3	3	2	2	2	2	2	2	27
25	3	3	2	1	1	3	2	3	2	1	2	1	3	3	3	3	3	1	1	0	0	0	0	0	14
26	3	3	2	1	2	1	3	1	3	2	3	2	1	3	3	1	0	0	0	0	0	0	0	0	7
27	3	3	2	1	3	2	1	2	1	3	1	3	2	3	3	3	3	3	3	3	3	3	3	3	33
Total														81	68	52	44	37	34	32	32	32	32	32	476

$\omega_1 = 100,000$ cycles, $\omega_2 = 200,000$ cycles, ... ,
$\omega_{10} = 1,000,000$ cycles, $\omega_{11} = 1,100,000$ cycles

At each level, a specimen was given a 1 if it was functioning and a 0 if
it had failed, so that a specimen that survived to level ω_{11} still received
only a 1. The data in Table 9.1 are the number of springs surviving in
each set of three. In Table 9.2 these are converted to 1/0 data. Then a
two-way array is prepared for A through F and ω to find the variations.

Table 9.2 1/0 Data

No.		ω_1	ω_2	ω_3	ω_4	ω_5	ω_6	ω_7	ω_8	ω_9	ω_{10}	ω_{11}	Total
	1	1	0	0	0	0	0	0	0	0	0	0	1
1	2	1	0	0	0	0	0	0	0	0	0	0	1
	3	1	0	0	0	0	0	0	0	0	0	0	1
	1	1	1	1	1	0	0	0	0	0	0	0	4
2	2	1	1	1	1	1	0	0	0	0	0	0	5
	3	1	1	1	1	1	1	1	1	1	1	1	11
	1	1	1	0	0	0	0	0	0	0	0	0	2
3	2	1	1	0	0	0	0	0	0	0	0	0	2
	3	1	1	1	1	1	1	1	1	1	1	1	11
⋮	⋮					⋮							⋮
	1	1	1	1	1	1	1	1	1	1	1	1	11
27	2	1	1	1	1	1	1	1	1	1	1	1	11
	3	1	1	1	1	1	1	1	1	1	1	1	11
Total		81	68	52	44	37	34	32	32	32	32	32	476

$$S_m = \frac{476^2}{27 \times 3 \times 11} = 254.29 \qquad ......(9.1)$$

$$S_T = 476 \qquad (f = 891) \qquad ......(9.2)$$

$$S_{T1} = \frac{1}{33}(3^2 + 20^2 + 15^2 + \cdots + 33^2) - S_m = 120.31 \qquad (f = 26) \qquad ......(9.3)$$

$$S_{T2} = \frac{1}{11}(1^2 + 1^2 + 1^2 + 4^2 + \cdots + 11^2) - S_m = 134.79 \; (f = 80) \qquad ......(9.4)$$

where S_{T1} is the variation among the 27 experiments and S_{T2} the
variation among the 81 springs.

$$S_A = \frac{120^2 + 169^2 + 187^2}{297} - 254.29 = 8.10 \quad (f = 2) \qquad \dots\dots(9.5)$$

$$S_{(BC)} = \frac{155^2 + 190^2 + 131^2}{297} - 254.29 = 5.92 \quad (f = 2) \qquad \dots\dots(9.6)$$

$$S_B = \frac{(155 - 131)^2}{594} = 0.97 \quad (f = 1) \qquad \dots\dots(9.7)$$

$$S_C = \frac{(155 - 190)^2}{594} = 2.06 \quad (f = 1) \qquad \dots\dots(9.8)$$

$$S_D = \frac{95^2 + 157^2 + 224^2}{297} - 254.29 = 28.03 \quad (f = 2) \qquad \dots\dots(9.9)$$

$$S_E = \frac{180^2 + 175^2 + 121^2}{297} - 254.29 = 7.21 \quad (f = 2) \qquad \dots\dots(9.10)$$

$$S_{D \times E} = \frac{38^2 + 38^2 + \cdots + 54^2}{99} - 254.29 - S_D - S_E = 5.50 \quad (f = 4) \qquad \dots\dots(9.11)$$

$$S_F = \frac{70^2 + 236^2 + 170^2}{297} - 254.29 = 47.04 \quad (f = 2) \qquad \dots\dots(9.12)$$

$$S_{D \times F} = \frac{10^2 + 38^2 + 47^2 + \cdots + 80^2}{99} - 254.29 - S_D - S_F = 12.60 \quad (f = 4) \qquad \dots\dots(9.13)$$

$$S_G = \frac{160^2 + 149^2 + 167^2}{297} - 254.29 = 0.55 \qquad \dots\dots(9.14)$$

$$S_{e1} = S_{T1} - (S_A + S_{(BC)} + \cdots + S_G) = 5.36 \quad (f = 6) \qquad \dots\dots(9.15)$$

$$S_{e2} = S_{T2} - S_{T1} = 134.79 - 120.31 = 14.48 \quad (f = 54) \qquad \dots\dots(9.16)$$

$$S_\omega = \frac{81^2 + 68^2 + \cdots + 32^2}{81} - 254.29 = 35.46 \quad (f = 10) \qquad \dots\dots(9.17)$$

$$S_{A \times \omega} = \frac{27^2 + 20^2 + \cdots + 14^2}{27} - S_m - S_A - S_\omega = 2.26 \quad (f = 20) \qquad \dots\dots(9.18)$$

$$\vdots$$

$$S_{G \times \omega} = \frac{27^2 + 22^2 + \cdots + 12^2}{27} - S_m - S_G - S_\omega = 1.10 \quad (f = 20)$$

$$S_{e3} = S_T - (S_m + S_{T2} + S_\omega + S_{A \times \omega} + \cdots + S_{G \times \omega}) = 19.98 \quad (f = 600) \qquad \dots\dots(9.19)$$

The above figures give the analysis of variance in Table 9.3. The factors can be rearranged as in Table 9.4 by pooling the factors designated by single circles with e_1 and the factors designated by double circles with e_2.

Table 9.3 Analysis of Variance

Factor	f	S	V	Factor	f	S	V
m	1	254.29	254.29	ω	10	35.46	3.55
A	2	8.10	4.05	$A \times \omega$	20	2.26	0.11◎
(BC)	2	5.93	2.96	$(BC) \times \omega$	20	2.47	0.12◎
(B)	(1)	0.97	0.97○	$(B) \times \omega$	(10)	1.29	0.13◎
(C)	(1)	2.06	2.06	$(C) \times \omega$	(10)	0.59	0.06◎
D	2	28.03	14.02	$D \times \omega$	20	4.00	0.20
E	2	7.21	3.60	$E \times \omega$	20	2.59	0.13◎
$D \times E$	4	5.50	1.38○	$D \times E \times \omega$	40	2.03	0.05◎
F	2	47.04	23.52	$F \times \omega$	20	5.80	0.29
$D \times F$	4	12.60	3.15	$D \times F \times \omega$	40	11.23	0.28
G	2	0.55	0.28○	$G \times \omega$	20	1.10	0.06◎
e_1	6	5.36	0.89○**	e_3	600	19.98	0.033
e_2	54	14.48	0.27**	T	891	476.00	

Table 9.4 Rearranged Analysis of Variance

Factor	f	S	V	S'	ρ (%)
m	1	254.29	254.29	253.34	53.22
A	2	8.10	4.05	6.20	1.30
C	1	4.96	4.96	4.01	0.84
D	2	28.03	14.02	26.13	5.49
E	2	7.21	3.60	5.31	1.12
F	2	47.04	23.52	45.14	9.48
$D \times F$	4	12.60	3.15	8.80	1.85
e_1	13	12.37	0.95	18.58	3.90
e_2	54	14.48	0.263	18.155	3.82
ω	10	35.46	3.546	35.037	7.33
$D \times \omega$	20	4.00	0.200	3.154	0.66
$F \times \omega$	20	5.80	0.290	4.954	1.04
$D \times F \times \omega$	40	11.23	0.281	9.538	2.00
e_3	720	30.43	0.0423	37.652	7.92
T	891	476.00		476.000	100.00

Clutch Spring Life Test: Estimation

From the analysis-of-variance table, we shall estimate the main effects of m, A, C, and E, and the three-way array for D, F, and ω.

$$\overline{T} = \frac{476}{891} = 0.534 \pm 0.071 \qquad\qquad\qquad ......(9.20)$$

$$\left. \begin{aligned} \overline{A_1} &= \frac{120}{297} = 0.404 \pm 0.122 \\[2em] \overline{A_2} &= \frac{169}{297} = 0.569 \pm 0.122 \\[2em] \overline{A_3} &= \frac{187}{297} = 0.630 \pm 0.122 \end{aligned} \right\} \qquad ......(9.21)$$

$$\left. \begin{aligned} \overline{C_1} &= \frac{286}{594} = 0.481 \pm 0.086 \\[2em] \overline{C_2} &= \frac{190}{297} = 0.640 \pm 0.122 \end{aligned} \right\} \qquad ......(9.22)$$

$$\left. \begin{aligned} \overline{E_1} &= \frac{180}{297} = 0.606 \pm 0.122 \\[2em] \overline{E_2} &= \frac{175}{297} = 0.589 \pm 0.122 \\[2em] \overline{E_3} &= \frac{121}{297} = 0.407 \pm 0.022 \end{aligned} \right\} \qquad ......(9.23)$$

The effects of A, C, and E (in which only the main effects are significant) are reliable, and so is the effect of DF, which has significant interaction with ω. Even at the worst levels of A, C, and E, with conditions D_2F_2 and D_3F_2, all the specimens last for 1,100,000 cycles, variations of the other factors notwithstanding.

The optimum conditions are $A_3C_2E_1D_2F_2$ or better one $A_3C_2E_1D_3F_2$. Since the survival rate with D_3F_2 or D_2F_2 is 100%, there is no need to calculate the process average. A confirmation test is all that is required. To illustrate the method, however, we shall estimate the process average for the present conditions $A_2\,C_2\,D_1\,E_3\,F_2$.

$$\hat{\mu} = \overline{D_1\,F_2\,\omega_i} + \overline{A_2} + \overline{C_2} + \overline{E_3} - 3\,\overline{T} \qquad \text{(db)}$$

Table 9.5   Three-Way Array for D, F, and ω
(Shown as totals. Divide by 9 to obtain means.)

	ω_1	ω_2	ω_3	ω_4	ω_5	ω_6	ω_7	ω_8	ω_9	ω_{10}	ω_{11}	Total	Survival rate
D_1F_1	9	1	0	0	0	0	0	0	0	0	0	10	0.101
D_1F_2	9	9	8	4	2	1	1	1	1	1	1	38	0.384
D_1F_3	9	7	5	4	4	3	3	3	3	3	3	47	0.474
D_2F_1	9	6	0	0	0	0	0	0	0	0	0	15	0.152
D_2F_2	9	9	9	9	9	9	9	9	9	9	9	99	1.000
D_2F_3	9	9	5	4	3	3	2	2	2	2	2	43	0.434
D_3F_1	9	9	7	5	3	2	2	2	2	2	2	45	0.455
D_3F_2	9	9	9	9	9	9	9	9	9	9	9	99	1.000
D_3F_3	9	9	9	9	7	7	6	6	6	6	6	80	0.808
Total	81	68	52	44	37	34	32	32	32	32	32	476	0.534
Survival rate	1.000	0.840	0.642	0.543	0.457	0.420	0.395	0.395	0.395	0.395	0.395		

To correct these values to decibels, Omega conversion tables in Appendix - 1 are used.

$$\overline{A}_2 = 0.569 \quad 1.21 \text{ decibels}$$
$$\overline{C}_2 = 0.640 \quad 2.50 \text{ decibels}$$
$$\overline{E}_3 = 0.407 \quad -1.63 \text{ decibels}$$
$$\overline{T} = 0.534 \quad 0.59 \text{ decibels}$$

The decibel values of $D_1F_2\,\omega_1$ through $D_1F_2\,\omega_{11}$ are:

$$1.0, \quad 1.0, \quad 0.89, \quad 0.44, \quad 0.22, \quad 0.11, \quad ..., \quad 0.11$$
$$\infty, \quad \infty, \quad 9.04, \quad -0.98, \quad -5.44, \quad -9.04, \quad ..., \quad -9.04$$

Calculating

$$A_2 + C_2 + E_3 - 3 \times T = 1.21 + 2.50 - 1.63 - 3 \times 0.59 = 0.31$$

and adding this to the above values gives the decibel values:

$$\infty, \quad \infty, \quad 9.35, \quad -0.67, \quad -5.13, \quad -8.73, \quad ..., \quad -8.73$$

The survival rates are therefore as shown in Table 9.6. See reference (1) for the calculation of the confidence limits.

Table 9.6   Estimated Process Averages (%)

	ω_1	ω_2	ω_3	ω_4	ω_5	ω_6	ω_7	ω_8	ω_9	ω_{10}	ω_{11}
	100.0	100.0	89.6	46.1	23.5	11.8	11.8	11.8	11.8	11.8	11.8
Lower confidence limit	96.0	96.0	72.9	20.1	8.7	4.0	4.0	4.0	4.0	4.0	4.0
Upper confidence limit	100.0	100.0	96.5	73.3	49.5	30.0	30.0	30.0	30.0	30.0	30.0

PROBLEM

To find the main effect of five factors A, B, C, D, and E involved in the manufacture of flourescent lamps and the $A \times B$ interaction, two levels were set for each factor, they were assigned to an L_8 array, and experiments were performed. The data shown below indicate the brightness (after the standard value of 40 watts is subtracted), and the life under certain stress conditions. Each experiment had two repetitions. In the life test, the lamps were checked every other day for 20 days to see if they were still functioning. A 4 would indicate that the lamp survived to day 4 but not to day 6. A double circle indicates a lamp that survived the full 20 days.

	A 1	B 2	$A \times B$ 3	C 4	D 5	E 6	e 7	Brightness		Life	
1	1	1	1	1	1	1	1	0	5	14	◎
2	1	1	1	2	2	2	2	3	4	20	◎
3	1	2	2	1	1	2	2	0	9	8	10
4	1	2	2	2	2	1	1	2	1	18	◎
5	2	1	2	1	2	1	2	2	2	◎	◎
6	2	1	2	2	1	2	1	−4	3	13	◎
7	2	2	2	1	2	2	1	−6	−1	17	◎
8	2	2	2	2	1	1	2	2	5	12	14

(1) Using levels of $\omega_1 = -6$, through $\omega_{16} = 9$ for the brightness and $\omega_1 = 0$, $\omega_2 = 2$, and so on to $\omega_{11} = 20$ for the life, perform a minute analysis.

(2) Using the significant factors, estimate the process average under the optimum conditions.

DISCUSSION

Standard Deviation and Decibel Values

G (Genichi): It is important to realize that most characteristic values should be expressed as S/N ratios or similar decibel values, but it is also permissible to compare distributions, as in this chapter. The idea is to transform the variables so as to increase the sensitivity to factor effect. If a characteristic value is nonnegative and varies by several orders of magnitude, it is important to convert to true data. For characteristics that in theory can take on arbitrary positive values but cannot be negative, see Chapters 7 and 8.

S (Student): So the *SN* ratio can be used not only to evaluate dynamic characteristics but also as a technique for transformation of variables, or to evaluate filter performance.

G: In some American writings, estimation theory is called the theory of filters. The role of a filter is to take a mixture of signal and noise, amplify only the signal, and attenuate the noise (relative to the signal). In a set of readings, factor effects are mixed with error effects. A good transformation of variables is one that amplifies the factor effects and attenuates the error effects.

S: But if a nonlinear transformation is used, won't nonsignificant factors become significant, leading to meaningless results?

G: If there is no signal factor effect to begin with, all you have is error factor effects. No matter how you transform the variables, the signal factor effect is not going to be significant. It's the same as when a radio station is not broadcasting. No matter what kind of filter you have in your receiver, you are not going to hear music. When you get music loud and clear, you know you have a receiver with a good filter. No transformation can create nonexistent signals. You needn't worry.

S: So if you do detect something by a transformation, it means you have increased the sensitivity.

G: That's right. I would prefer to have no interactions, but I think interactions can be treated as a kind of error.

10 EXPERIMENTAL DESIGN FOR DYNAMIC CHARACTERISTICS

Dynamic Characteristics

Among the different types of design study, orthogonal arrays and analysis of variance are particularly useful in dealing with dynamic characteristics. Typical examples of dynamic characteristics are the velocity characteristics (speed and direction) of an automobile. If the goal is to design an automobile that the driver can easily control, how should the necessary experiments be designed?

Let us begin by defining dynamic characteristics. The controllability of an automobile or airplane, the performance of a machine tool, robot, or tennis racket, and the athletic abilities of a human being are all examples of dynamic characteristics. Dynamic characteristics exhibit the following pattern.

Intention or target ⟶ Signal ⟶ Result

In skiing, for example, the skier positions his center of gravity on the midline or to the right depending on whether he wants to go straight or wants to turn right. Given an intention or target, a control variable (signal factor) must be altered to carry it out. When there are many possible intentions, such as going straight or turning by various degrees to the right, the signal factor (the skier's center of gravity) must be shifted by an amount that matches the target. This is the reason for the term dynamic.

Dynamic performance may not be necessary. For instance, the straight- ahead characteristics of jump skis can be improved by making grooves on their bottom surfaces because the skier never needs to turn in a jump competition. Suppose, however, that you are working with sheet metal and want good rollability characteristics: that is, you want metal that can be rolled to a variety of thicknesses, but without variation once the thickness is decided, so that if you roll for a thickness of 1 mm, then 1 mm is what you get, and if you roll for a thickness of 0.8 mm, then 0.8 mm is what you get. A product that can meet a variety of different requirements like this is said to have good dynamic characteristics.

The different requirements -- 1 mm thickness this time, 0.8 mm thickness the next time -- are satisfied by changing the levels of signal factors, in this case the rolling pressure and the number of times the metal is rolled.

The signal factors that control a skier's turn are his shifting of his center of gravity and his use of the edges of the skis. For an automobile, the signal factor controlling the direction of travel is the steering angle -- the degree of rotation of the steering wheel -- and speed is controlled by the accelerator and brake pedals. For the same amount of steering to the right or left, however, the turning radius will vary depending on the road conditions and type of tires. If the effect of the signal factors is certain, but the direction of travel varies due to other factors (error factors), the automobile is unstable. An automobile should be designed for reliable signal factor effect and minimal error factor effect.

A dye is said to have good dyeing characteristics if it is possible to obtain any desired degree of saturation of the color. If the saturation can be changed by adjusting the quantity of dye, the quantity is the signal factor. If the saturation varies with the same quantity of dye because of thread thickness, thread count in the cloth, temperature, or agitation during the dyeing process, these are not good dyeing characteristics; the variations are caused by non-signal factors. Good dyeing characteristics mean that the saturation varies consistently with the quantity of dye, and is affected as little as possible by other factors (error factors or noise factors). This means a high S/N ratio, the S/N ratio being the size of the signal factor effect divided by the size of the error factor effect.

It is possible for the effect of a signal factor to be too large -- for the sensitivity to be too high. If a small change in the quantity of dye causes a large change in the color saturation, the dye will be hard to use. The dyer will have to measure extremely carefully; the smallest mistake will be fatal. A dyeing characteristic such as ease of use cannot be measured by the S/N ratio alone.

Let M stand for the signal factor (quantity of dye). If a small change in M produces a large change in the saturation y of the color, this means that the function

$$y = f(M)$$

has a large derivative

$$y' = f'(M)$$

Sensitivity is defined as the square of this derivative, $f'(M)^2$. Fortunately, oversensitivity is a problem that can always be corrected. If the dye is diluted by an appropriate additive, for example, the sensitivity can be reduced to any desired degree. Accordingly, the essential goal in quality improvement is to improve the S/N ratio. The sensitivity can be adjusted afterward.

Power supply circuits and similar devices must provide immediate output in response to signals or input, so their performance can be regarded as dynamic. In studying the power circuit of a television set that converts 100 V AC to 115 V DC, for example, stability is not the only characteristic that needs to be considered. It would be one thing if all of the circuits produced at the factory delivered precisely 115-V output in response to 100-V input, but no parameter design can reduce the production variability to zero. If the output specification is 115 ± 1.4 V and a circuit outputs only 112 V, it is defective. The question is closed if the circuit is scrapped, but in many cases it will be adjusted to the 115-V target by replacing a component such as a resistor and shipped. If the output voltage of the circuit did not change in response to a change in the value of any component characteristic, adjustment would be impossible. There has to be an adjustment variable with a sufficient effect. Here the variable used to correct the difference from the target of 115 V is the adjustment factor, and the adjustment factor has to be sufficiently sensitive to have an effect.

A similar situation is found in machine tools. When a product is machined by a machine tool, it is desirable that the deviation be correctable as efficiently as possible if a characteristic value deviates from its target. If there are three objective characteristics, at least three independent variables (adjustment factors) are needed to adjust them. If possible, the effect of each control factor should be limited as much as possible to one objective characteristic. The designer should therefore assign one adjustment factor to each objective characteristic and attempt to reduce the effect of other adjustment factors. Then if a target value is missed, it will be easy to calculate the amount of correction necessary.

Automatic control and automatic adjustment mechanisms are the embodiment of dynamic characteristics. The variable that corrects deviation from the target value is the signal factor.

Not many experiments on dynamic characteristics have been published, but one valuable reference is *S/N-hi ni yoru dainamikkuna tokusei no hyoka* (Evaluation of dynamic characteristics by *S/N* ratio), a report by the *S/N* Ratio Working Group of the JSA Control Systems Committee. We shall borrow an example from that source.

Experiment on Truck Steerability

The driving characteristics of a vehicle are what enable the driver to control its velocity (speed and direction). The experiment described below was carried out by a design group at Isuzu Motors in 1974 to investigate directional driving characteristics, that is, steerability.

The purposes of this experiment were to learn how much steerability would suffer if the usual Ackerman steering geometry were replaced by parallel geometry, and to determine the optimum hardness of the front springs in relation to the steering geometry. (In the Ackerman geometry, the inside and outside wheels are cut at different angles to give a smooth circular turn. The parallel geometry is a link system in which the inside and outside wheels remain parallel, so that there is a tendency during the turn for the tires to twist and the vehicle to slip sideways.) It was thought that changing from the current standard spring hardness might adequately prevent steerability deterioration even if the parallel system were used. Accordingly, there were two control variables.

(1) Control variables
    A: front springs A_1 soft
                             A_2 standard
                             A_3 hard
    B: steering geometry B_1 Ackerman
                             B_2 parallel

Next, the following three levels were selected as being typical speeds at which a car might turn. Speed was treated as an indicative factor -- a factor that is not controllable but for which a separate S/N ratio is desired for each level.

(2) Indicative factor
    C: speed C_1 15 km/h
                       C_2 20 km/h
                       C_3 25 km/h

Since there were 3×2 combinations of control factor levels and 3 indicative factor levels, there were only $3 \times 2 \times 3 = 18$ combinations to experiment with, and the full three-way layout was used. If there had been more control and indicative factors, they would have been assigned to an orthogonal array.

(3) Signal factor
    M: steering angle M_1 200°
                             M_2 250°
                             M_3 300°

(4) Error factors

K:	turning direction	K_1		right
		K_2		left
L:	road condition	L_1		wet asphalt
		L_2		dry asphalt
		L_3		dry concrete
N:	position of load	N_1		front
		N_2		standard
		N_3		back
P:	type of tire	P_1		rig tire
		P_2		rag tire
		P_3		radial tire
Q:	front tire air pressure	Q_1		6 kg/cm^2 right and left
		Q_2		6 kg/cm^2 right, 3 kg/cm^2 left
		Q_3		3 kg/cm^2 right and left

The signal and error factors were assigned to an outer L_{18} orthogonal array as shown in Table 10.1.

Table 10.1 Layout of Signal and Error Factors (Outer Orthogonal Array)

No.	K 1	L 2	M 3	e 4	N 5	P 6	Q 7	e 8
1	1	1	1	1	1	1	1	1
2	1	1	2	2	2	2	2	2
3	1	1	3	3	3	3	3	3
4	1	2	1	1	2	2	3	3
5	1	2	2	2	3	3	1	1
6	1	2	3	3	1	1	2	2
7	1	3	1	2	1	3	2	3
8	1	3	2	3	2	1	3	1
9	1	3	3	1	3	2	1	2
10	2	1	1	3	3	2	2	1
11	2	1	2	1	1	3	3	2
12	2	1	3	2	2	1	1	3
13	2	2	1	2	3	1	3	2
14	2	2	2	3	1	2	1	3
15	2	2	3	1	2	3	2	1
16	2	3	1	3	2	3	1	2
17	2	3	2	1	3	1	2	3
18	2	3	3	2	1	2	3	1

The turning radius was measured under the conditions of the outer L_{18} array for all 18 combinations of the inner factors A, B, and C. Two trucks, B_1 and B_2, were used. The 54 combinations of the three levels of front spring hardness (A_1, A_2, and A_3) with the outer L_{18} array were tested in the most convenient order. For each combination, the speed was varied at three levels and the radius in which the truck made full circle was measured. The resulting data are given in Table 10.2.

Table 10.2 Turning Radius
(In meters, medium speed, A_1B_1 and A_1B_2 only, rest omitted)

No.	A_1B_1			A_1B_2		
	C_1 (15km/h)	C_2 (20km/h)	C_3 (25km/h)	C_1 (15km/h)	C_2 (20km/h)	C_3 (25km/h)
1	41.9	44.0	47.6	40.3	41.7	44.3
2	33.7	34.2	34.9	28.3	29.3	30.3
3	27.2	27.4	26.7	23.9	22.8	22.8
4	42.2	43.7	45.2	39.1	40.3	41.7
5	31.4	31.9	31.8	29.2	29.4	29.0
6	30.1	31.8	34.5	26.4	29.8	32.4
7	44.4	46.6	49.4	43.0	46.2	48.9
8	33.6	35.5	37.2	32.3	33.4	35.0
9	27.0	27.3	26.9	24.7	24.7	24.4
10	37.4	37.2	37.1	40.4	40.4	41.0
11	36.2	42.8	45.0	37.3	43.0	46.6
12	25.5	25.5	26.7	25.9	26.3	26.1
13	38.1	38.2	37.0	37.0	36.8	36.7
14	31.8	33.7	36.2	34.0	35.8	38.5
15	26.7	27.6	29.2	26.3	27.4	32.3
16	38.1	39.5	40.6	35.7	37.3	38.8
17	27.6	28.1	27.5	28.5	29.1	28.5
18	30.2	34.6	39.0	29.3	33.7	37.4

For the high-speed steerability test, there is not enough space for the truck to make a full circle, so the turning radius will be measured by having the truck turn through a quarter circle or less, taking three points A, B, and C on its path, and finding the intersection of the perpendicular bisectors of the segments AB and BC.

Calculation of S/N Ratio

For each of the 18 combinations in the inner layout, the S/N ratio was calculated from the data for the outer array using the formula:

$$\eta = \frac{\dfrac{1}{r \cdot s \cdot h^2}(S_\beta - V_e)}{V_e} \qquad \qquad ......(10.1)$$

Experimental Design for Dynamic Characteristics

where h was measured in units of 50°, so that for 50°, $h = 1$. The numerator β^2 of the S/N ratio η is the amount of change in the turning radius when the steering angle was changed by a unit of 50°. The reason for using units of 50° was so that there would not be a long string of zeros after the decimal point in the value of the S/N ratio η. The calculations are shown in Tables 10.3 and 10.4. They were calculated by computer with a higher precision than indicated.

Table 10.3 Calculation of S/N Ratios

A B C	S_T	S_β	S_e	V_e	η	Decibel
111	586.67	473.76	112.91	7.06	5.51	7.41
112	719.67	468.75	250.92	15.68	2.41	3.82
113	907.86	455.10	452.76	28.30	1.26	0.99
121	641.02	520.08	120.94	7.56	5.65	7.52
122	790.70	507.00	283.70	17.73	2.30	3.62
123	984.06	481.33	502.73	31.42	1.19	0.77
211	671.16	500.52	170.64	10.66	3.93	5.94
212	870.09	501.81	368.28	23.02	2.06	3.14
213	1088.50	517.45	571.05	35.69	1.30	1.13
221	661.96	517.45	144.51	9.03	4.80	6.81
222	707.84	471.25	236.59	14.79	2.57	4.10
223	904.00	453.87	450.13	28.13	1.48	1.69
311	665.29	537.34	127.95	8.00	5.51	7.41
312	796.04	514.83	281.21	17.58	2.82	4.50
313	944.50	471.25	473.25	29.58	1.48	1.70
321	735.28	620.64	114.64	7.16	7.26	8.61
322	871.06	558.97	312.09	19.51	2.91	4.64
323	1159.01	512.21	646.80	40.42	1.22	0.85

Analysis of S/N Ratio

In this example the S/N ratios are measures of steerability, so the S/N ratio (decibel value) is analyzed as an objective characteristic with respect to the inner factors. The decibel value must always be used for the S/N ratio. Since the inner factors A, B, and C are assigned to a three-way layout, the S/N ratios are obtained for all 18 of their combinations, and Table 10.4 can be analyzed as follows.

First supplementary layouts are constructed as in Table 10.5.
From Tables 10.4 and 10.5, the variations of each source of factorial effects can be calculated as follows.

$$S_T = 7.41^2 + 3.82^2 + \cdots + 0.85^2 - \frac{74.65^2}{18}$$

$$= 118.0328 \qquad (f = 17) \qquad \qquad \dots\dots(10.2)$$

$$S_A = \frac{24.13^2 + 22.81^2 + 27.71^2}{6} - \frac{74.65^2}{18}$$

$$= 2.1427 \qquad (f = 2) \qquad \qquad \dots\dots(10.3)$$

$$S_B = \frac{36.04^2 + 38.61^2}{9} - \frac{74.65^2}{18}$$

$$= 0.3669 \qquad (f = 1) \qquad \qquad \dots\dots(10.4)$$

$$S_{A \times B} = \frac{12.22^2 + 11.91^2 + \cdots + 14.10^2}{3} - \frac{74.65^2}{18} - S_A - S_B$$

$$= 0.6412 \qquad (f = 2) \qquad \qquad \dots\dots(10.5)$$

$$S_C = \frac{43.70^2 + 23.82^2 + 7.13^2}{6} - \frac{74.65^2}{18}$$

$$= 111.7297 \qquad (f = 2) \qquad \qquad \dots\dots(10.6)$$

$$S_{A \times C} = \frac{14.93^2 + 7.44^2 + \cdots + 2.55^2}{2} - \frac{74.65^2}{18} - S_A - S_C$$

$$= 2.0230 \qquad \qquad \dots\dots(10.7)$$

$$S_{B \times C} = \frac{20.76^2 + 11.46^2 + \cdots + 3.31^2}{3} - \frac{74.65^2}{18} - S_B - S_C$$

$$= 0.6036 \qquad \qquad \dots\dots(10.8)$$

$$S_e = S_T - (S_A + S_B + S_C + S_{A \times B} + S_{A \times C} + S_{B \times C})$$

$$= 0.5257 \qquad \qquad \dots\dots(10.9)$$

This gives the analysis of variance shown in Table 10.6.

The results of analysis of variance indicate that spring hardness A is significant at the 5% level and speed C is significant at the 1% level. The significant factor effects are estimated below.

$$\overline{A_1} = \frac{24.13}{6} = 4.02 \pm 0.45 \qquad \qquad \dots\dots(10.10)$$

$$\overline{A_2} = \frac{22.81}{6} = 3.80 \pm 0.45 \qquad \qquad \dots\dots(10.11)$$

Table 10.4 Layout of Inner Factors and Data (Three-Way Layout)

	A	B	C	Decibel
1	1	1	1	7.41
2	1	1	2	3.82
3	1	1	3	0.99
4	1	2	1	7.52
5	1	2	2	3.62
6	1	2	3	0.77
7	2	1	1	5.94
8	2	1	2	3.14
9	2	1	3	1.13
10	2	2	1	6.81
11	2	2	2	4.10
12	2	2	3	1.69
13	3	1	1	7.41
14	3	1	2	4.50
15	3	1	3	1.70
16	3	2	1	8.61
17	3	2	2	4.64
18	3	2	3	0.85
			計	74.65

Table 10.5 Supplementary Layouts

$r = 3$	B_1	B_2	Total
A_1	12.22	11.91	24.13
A_2	10.21	12.60	22.81
A_3	13.61	14.10	27.71
Total	36.04	38.61	74.65

$r = 2$	C_1	C_2	C_3	Total
A_1	14.93	7.44	1.76	24.13
A_2	12.75	7.24	2.82	22.81
A_3	16.02	9.14	2.55	27.71
Total	43.70	23.82	7.13	74.65

$r = 3$	C_1	C_2	C_3	Total
B_1	20.76	11.46	3.82	36.04
B_2	22.94	12.36	3.31	38.61
Total	43.70	23.82	7.13	74.65

Table 10.6 Analysis of Variance of S/N Ratios

Source	f	S	V	F_0
A	2	2.1427	1.0714	4.52*
B	1	0.3669	0.3669	○
C	2	111.7297	55.8648	235.62**
$A \times B$	2	0.6412	0.3206	○
$A \times C$	4	2.0230	0.5058	2.13△
$B \times C$	2	0.6036	0.3018	○
e	4	0.5257	0.1314	○
(e)	(9)	(2.1344)	(0.2371)	
T	17	118.0328		

Pooled from circled sources

$$\overline{A_3} = \frac{27.71}{6} = 4.62 \pm 0.45 \qquad\qquad ......(10.12)$$

$$\overline{C_1} = \frac{43.70}{6} = 7.28 \pm 0.45 \qquad\qquad ......(10.13)$$

$$\overline{C_2} = \frac{23.82}{6} = 3.97 \pm 0.45 \qquad\qquad ......(10.14)$$

$$\overline{C_3} = \frac{7.13}{6} = 1.19 \pm 0.45 \qquad\qquad ......(10.15)$$

The confidence limits are calculated from the following formula:

$$\pm \sqrt{\frac{5.12 \times 0.2371}{6}} = \pm 0.45 \qquad\qquad ......(10.16)$$

These results are graphed in Figure 10.1.

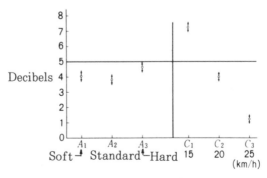

Figure 10.1 Graph of Significant Factors

A_3 is the optimum spring hardness. The difference in steering geometry is not significant. The effect of speed is unexpectedly large. Although as can be seen from S_β in Table 10.3 the effect of the steering angle remains fairly constant, as the speed increases, the effects of error factors severely degrade the S/N ratio: by 3.3 decibels between C_1 and C_2, and by 6.1 decibels between C_1 and C_3. This indicates that turning at high speed is dangerous due to poor steerability. Poor steerability can lead to accidents, but we shall forgo a calculation of the loss function.

PROBLEM

To find conditions that would reduce the variability of the diameter of holes made by a cutting tool, the following factors were studied:

$A =$	Cutting speed (m/min)	28	24
$B =$	Feed rate (m/rev)	0.016	0.011
$C =$	Type of bit	present	proposed
$D =$	Bite (mm)	0.05	0.5
$E =$	Guide bush material	present	proposed
$F =$	Guide bush tension	strong	weak
$G =$	Amount of lubricant	present	double

These were assigned to an L_{16} orthogonal array as follows to obtain the main effects and the $A \times C$, $D \times F$, and $E \times G$ interactions. For each set of conditions, three specimens were cut from two lots each and their diameters measured. For experiment number 1, the following data were obtained (in units of 0.01 mm):

34 37 40 28 32 25

The following S/N ratio was then obtained as a measure of the size of the variability.

$$\eta = 10\log \frac{\frac{1}{6}(S_m - V_e)}{V_e} = 10\log \frac{\frac{1}{6}(6402.7 - 31.1)}{31.1}$$

$$= 15.3 \text{ decibels}$$

Where

$$S_m = \frac{(34 + 37 + \cdots + 25)^2}{6} = 6402.7$$

$$V_e = \frac{1}{5}[34^2 + 37^2 + \cdots + 25^2 - S_m] = 31.1$$

The table below shows the S/N ratios obtained in the same way for the other experiments.
(1) Make an analysis-of-variance table.
(2) Estimate the significant factors.
(3) Find the optimum conditions and estimate the process average.
(4) If the target value of the diameter is $300 \pm 100\mu m$, what will be the loss due to variability under the optimum conditions? The loss from a defective product is ¥ 30.

Layout and Values of S/N Ratio

	F 1	D 2	$F \times D$ 3	C 4	e 5	G 6	e 7	$B \times D$ 8	$A \times C$ 9	B 10	e 11	E 12	A 13	e 14	e 15	Data (decibels)
1	1	1	1	1	1	1	1	1	1	1	1	1	1	1	1	15.3
2	1	1	1	1	1	1	1	2	2	2	2	2	2	2	2	18.2
3	1	1	1	2	2	2	2	1	1	1	1	2	2	2	2	22.0
4	1	1	1	2	2	2	2	2	2	2	2	1	1	1	1	21.3
5	1	2	2	1	1	2	2	1	1	2	2	1	1	2	2	13.5
6	1	2	2	1	1	2	2	2	2	1	1	2	2	1	1	14.4
7	1	2	2	2	2	1	1	1	1	2	2	2	2	1	1	20.0
8	1	2	2	2	2	1	1	2	2	1	1	1	1	2	2	17.2
9	2	1	2	1	2	1	2	1	2	1	2	1	2	1	2	18.5
10	2	1	2	1	2	1	2	2	1	2	1	2	1	2	1	16.1
11	2	1	2	2	1	2	1	1	2	1	2	2	1	2	1	18.9
12	2	1	2	2	1	2	1	2	1	2	1	1	2	1	2	23.2
13	2	2	1	1	2	2	1	1	2	2	1	1	2	2	1	20.3
14	2	2	1	1	2	2	1	2	1	1	2	2	1	1	2	14.1
15	2	2	1	2	1	1	2	1	2	2	1	2	1	1	2	19.4
16	2	2	1	2	1	1	2	2	1	1	2	1	2	2	1	23.3

DISCUSSION

Experimental Results

S (Student):In the analysis of variance by SN ratio in Table 10.6, it turned out there was no significant difference between the two steering geometries, despite the fact that one would naturally expect the Ackerman geometry (B_1) to be better than the parallel geometry (B_2). What should this mean to an engineer?

G (Genichi):The fact that there is no significant difference does not mean that there is no difference at all. It just means that any difference there may be is small. Try finding the difference between the two levels of factor B and calculating the confidence limits.

S: All right, here it is:

$$\bar{B}_1 - \bar{B}_2 \pm \sqrt{F \times V_e \times \frac{2}{9}}$$

$$= \frac{36.04}{9} - \frac{38.61}{9} \pm \sqrt{5.12 \times 0.2371 \times \frac{2}{9}}$$

$$= -0.28 \pm 0.52 \qquad\qquad\qquad ......(D\text{-}10.1)$$

This seems to say that B_1 might be 0.24 decibels better than B_2, or it might be 0.80 decibels worse.

G: As was stated in the text, the Isuzu design group performed this experiment to show that there would not be a large difference in steerability between B_1 and B_2, so when they found no significant difference, they got what they were expecting. In experiments at lower speeds (omitted in this book), however, the Ackerman system did show a significantly better S/N ratio.

S: I've seen those data. At the lower speeds, they were turning the steering wheel about 700°, to nearly its maximum limit, so they were probably straining the capabilities of the parallel steering geometry beyond what it was intended to tolerate. That is why the Ackerman system came out better.

G: True. As mentioned in the first paragraph of this chapter, this example is one of the few that have been published, and it is also quoted in reference (1). The lower- speed experiments described there were performed in exactly the same pattern as in Tables 10.1 and 10.2, but with the three levels of the signal factor M and the speed C being:

$M_1 = 560°$ $C_1 = 5$ km/h
$M_2 = 700°$ $C_2 = 10$ km/h
$M_3 = 840°$ $C_3 = 15$ km/h

S: They used the same signal and error factors as in Tables 10.1 and 10.2, and they generated twice as much data as shown in Table 10.2. How long did the experiments take?

G: Using one Ackerman-type truck and one parallel-type truck, the tests were apparently completed in two days -- not counting preparation time.

S: To measure the turning radius, all you have to do is drive the truck in a circle. The time must have been taken by changing the front springs, changing the tires, and altering the road conditions.

G: Even if the experiments were performed in the most convenient order, there were $36 \times 18 = 648$ data values to be found, which must have meant a lot of work. The levels of the control factors and indicative factors were set as follows and assigned to a four-way inner layout.

$A =$ front spring hardness   three levels
$B =$ steering geometry   two levels
$M_1' =$ medium speed, $M_2' =$ low speed   two levels
$C =$ speed   three levels

S: For the speed C there were three levels for M_1' (15, 20, and 25 km/h), and three levels for M_2' (5, 10, and 15 km/h). That makes 36 sets of conditions in the inner layout alone.

G: Perhaps they should have used an L_{18} or L_{36} array and assigned more control factors. Then they might have found out more about steerability.

S: The type of tire, for example, is important in improving steerability, so shouldn't that be a control factor?

G: From the viewpoint of the tire manufacturer, it should. The vehicle designer, however, wants to be able to say that the vehicle has good steerability and will not slip regardless of what kind of tires are used, or whether they are somewhat worn, or whether the air pressure is uneven. For him, the tire difference is an error factor.

S: And the tire designer wants to be able to say that his tires have good steerability and will not slip no matter how the vehicle is designed or what the road conditions are, so for him the difference between vehicles is an error factor.

G: True. Here are the conclusions drawn by the Isuzu design group, quoted from chapter 24 in reference (1):

(1) Effect of steering angle: M_1' and M_2' represent a fairly gentle turn and a fairly sharp turn, respectively. The analysis results indicate that the S/N ratio is higher at steering angles in the neighborhood of 250° than in the neighborhood of 700°. A high S/N ratio means that the five error factors cited in the experiment (turning direction, road condition, position of load, type of tire, and front tire air pressure) and other error factors not explicitly considered are less likely to affect steerability. A detailed analysis of the numerator and denominator of the S/N ratio shows that M_1' is nearly 6.3 times as sensitive as M_2', but with only 1.5 times the error factor effect.

(2) Effect of driving speed: In the case of both M_1' and M_2', the lower the speed, the higher the S/N ratio. If the numerator and denominator of the S/N ratio are compared separately, it turns out that as the speed increases, the size of the signal effect does not change very much but error effect becomes much larger.

(3) Steering geometry: At small steering angles (around 250°), the effect of the control factors is only 1.0 to 2.0 decibels or a little less.

With this large a turning radius, at these speeds, the vehicle is apparently stable enough that differences in the specifications show no noticeable effect. At large steering angles (around 700°), however, effects on the order of 3.5 to 4.0 decibels were observed. The highest combination was A_3B_1, and the lowest was A_3B_2. The difference is caused by the steering geometry.

B_1 is the Ackerman geometry, and B_2 is the parallel geometry. Briefly, the Ackerman geometry is designed to reduce directional tire slip and give a smooth, circular turn. In the parallel system the inside and outside tires are cut at the same angle; such that the tires are twisted, and slipping is frequent during a turn. However, the parallel geometry has other advantages, and is not necessarily inferior. In these tests, the behavior of the tires with respect to road conditions at low speeds was less stable in the parallel geometry than in the Ackerman geometry, which can be interpreted as meaning that the parallel geometry is more susceptible to the effects of road conditions, load, and air pressure. These conclusions are supported by past experience and theory.

(4) Front springs: The front spring effect (A) appears as an interaction with the steering geometry (B). Observing the relation between S/N ratio and vehicle speed at steering angles of about 700°, we find that $A_3 B_1$ gives the highest S/N ratio, and A_3B_2 the lowest, with A_1B_1 and $A_1 B_2$ in the middle. When the front springs are soft, there is no difference between the two geometries, but when the springs are hard, the Ackerman geometry is superior.

S: I begin to see the value of the S/N method here.

G: There are many other dynamic characteristics. Examples from digital systems and chemical systems are given in Chapters 23 and 24 of reference (1), and there are examples from measuring instruments and sensors in Chapter 22.

REFERENCES

(1) G. Taguchi, *System of Experimental Design* (2 volumes), UNIPUB/Kraus International Publications, New York, 1987.

(2) G. Taguchi, *On-Line Quality Control,* Japanese Standards Associations, Tokyo, 1986.

(3) G. Taguchi, *Off-Line Quality Control,* Central Japan Quality Control Association, Nagoya, 1980.

(4) Instrumental Control Committee, *Parameter Design for New Products,* Japanese Standards Association, 1984, To be published by UNIPUB, New York.

(5) Japanese Industrial Standard K7109, *General Tolerancing Rules for Plastics Dimensions,* Japanese Standards Association, Tokyo, 1986.

(6) G. Taguchi, E. Elsayed, T. Hsiang, *Quality Engineering in Production Systems,* To be published by McGraw-Hill, New York.

APPENDICES

APPENDIX 1. OMEGA CONVERSION TABLE

$\frac{P}{(\%)}$	db	$\frac{P}{(\%)}$	db	$\frac{P}{(\%)}$	db	$\frac{P}{(\%)}$	db	$\frac{P}{(\%)}$	db	$\frac{P}{(\%)}$	db
0.0	$-\infty$*	5.0	-12.787	10.0	-9.541	15.0	-7.532	20.0	-6.020	25.0	-4.770
0.1	-29.995	5.1	-12.696	10.1	-9.498	15.1	-7.498	20.1	-5.993	25.1	-4.747
0.2	-26.980	5.2	-12.607	10.2	-9.446	15.2	-7.465	20.2	-5.966	25.2	-4.724
0.3	-25.215	5.3	-12.520	10.3	-9.399	15.3	-7.431	20.3	-5.939	25.3	-4.701
0.4	-23.961	5.4	-12.434	10.4	-9.352	15.4	-7.397	20.4	-5.912	25.4	-4.678
0.5	-22.988	5.5	-12.350	10.5	-9.305	15.5	-7.364	20.5	-5.885	25.5	-4.655
0.6	-22.191	5.6	-12.267	10.6	-9.259	15.6	-7.331	20.6	-5.859	25.6	-4.632
0.7	-21.518	5.7	-12.185	10.7	-9.214	15.7	-7.298	20.7	-5.832	25.7	-4.610
0.8	-20.933	5.8	-12.105	10.8	-9.168	15.8	-7.266	20.8	-5.806	25.8	-4.587
0.9	-20.417	5.9	-12.026	10.9	-9.124	15.9	-7.233	20.9	-5.779	25.9	-4.564
1.0	-19.955	6.0	-11.949	11.0	-9.079	16.0	-7.201	21.0	-5.753	26.0	-4.542
1.1	-19.537	6.1	-11.872	11.1	-9.035	16.1	-7.168	21.1	-5.727	26.1	-4.519
1.2	-19.155	6.2	-11.797	11.2	-8.991	16.2	-7.136	21.2	-5.701	26.2	-4.497
1.3	-18.803	6.3	-11.723	11.3	-8.947	16.3	-7.104	21.3	-5.675	26.3	-4.474
1.4	-18.476	6.4	-11.650	11.4	-8.904	16.4	-7.073	21.4	-5.649	26.4	-4.452
1.5	-18.172	6.5	-11.578	11.5	-8.861	16.5	-7.041	21.5	-5.623	26.5	-4.429
1.6	-17.888	6.6	-11.507	11.6	-8.819	16.6	-7.010	21.6	-5.598	26.6	-4.407
1.7	-17.620	6.7	-11.437	11.7	-8.777	16.7	-6.978	21.7	-5.572	26.7	-4.385
1.8	-17.367	6.8	-11.368	11.8	-8.735	16.8	-6.947	21.8	-5.547	26.8	-4.363
1.9	-17.128	6.9	-11.300	11.9	-8.693	16.9	-6.916	21.9	-5.521	26.9	-4.341
2.0	-16.901	7.0	-11.233	12.0	-8.652	17.0	-6.885	22.0	-5.496	27.0	-4.319
2.1	-16.685	7.1	-11.167	12.1	-8.611	17.1	-6.855	22.1	-5.470	27.1	-4.297
2.2	-16.478	7.2	-11.101	12.2	-8.570	17.2	-6.824	22.2	-5.445	27.2	-4.275
2.3	-16.281	7.3	-11.037	12.3	-8.530	17.3	-6.794	22.3	-5.420	27.3	-4.253
2.4	-16.091	7.4	-10.973	12.4	-8.490	17.4	-6.763	22.4	-5.395	27.4	-4.231
2.5	-15.910	7.5	-10.910	12.5	-8.450	17.5	-6.733	22.5	-5.370	27.5	-4.209
2.6	-15.735	7.6	-10.848	12.6	-8.410	17.6	-6.703	22.6	-5.345	27.6	-4.187
2.7	-15.566	7.7	-10.786	12.7	-8.371	17.7	-6.673	22.7	-5.321	27.7	-4.166
2.8	-15.404	7.8	-10.725	12.8	-8.332	17.8	-6.644	22.8	-5.296	27.8	-4.144
2.9	-15.247	7.9	-10.665	12.9	-8.293	17.9	-6.614	22.9	-5.271	27.9	-4.122
3.0	-15.096	8.0	-10.606	13.0	-8.255	18.0	-6.584	23.0	-5.427	28.0	-4.101
3.1	-14.949	8.1	-10.547	13.1	-8.216	18.1	-6.555	23.1	-5.222	28.1	-4.079
3.2	-14.806	8.2	-10.489	13.2	-8.178	18.2	-6.526	23.2	-5.198	28.2	-4.058
3.3	-14.668	8.3	-10.432	13.3	-8.141	18.3	-6.497	23.3	-5.173	28.3	-4.036
3.4	-14.534	8.4	-10.375	13.4	-8.103	18.4	-6.468	23.4	-5.149	28.4	-4.015
3.5	-14.404	8.5	-10.319	13.5	-8.066	18.5	-6.439	23.5	-5.125	28.5	-3.994
3.6	-14.227	8.6	-10.263	13.6	-8.029	18.6	-6.410	23.6	-5.101	28.6	-3.972
3.7	-14.153	8.7	-10.209	13.7	-7.992	18.7	-6.381	23.7	-5.077	28.7	-3.951
3.8	-14.033	8.8	-10.154	13.8	-7.955	18.8	-6.353	23.8	-5.053	28.8	-3.930
3.9	-13.916	8.9	-10.100	13.9	-7.919	18.9	-6.325	23.9	-5.029	28.9	-3.909
4.0	-13.801	9.0	-10.047	14.0	-7.883	19.0	-6.296	24.0	-5.005	29.0	-3.888
4.1	-13.689	9.1	-9.994	14.1	-7.847	19.1	-6.268	24.1	-4.981	29.1	-3.867
4.2	-13.580	9.2	-9.942	14.2	-7.811	19.2	-6.240	24.2	-4.958	29.2	-3.846
4.3	-13.473	9.3	-9.890	14.3	-7.775	19.3	-6.212	24.3	-4.934	29.3	-3.825
4.4	-13.369	9.4	-9.839	14.4	-7.740	19.4	-6.184	24.4	-4.910	29.4	-3.804
4.5	-13.267	9.5	-9.788	14.5	-7.705	19.5	-6.157	24.5	-4.887	29.5	-3.783
4.6	-13.167	9.6	-9.738	14.6	-7.670	19.6	-6.129	24.6	-4.863	29.6	-3.762
4.7	-13.069	9.7	-9.688	14.7	-7.635	19.7	-6.101	24.7	-4.840	29.7	-3.741
4.8	-12.973	9.8	-9.639	14.8	-7.601	19.8	-6.074	24.8	-4.817	29.8	-3.720
4.9	-12.879	9.9	-9.590	14.9	-7.566	19.9	-6.047	24.9	-4.793	29.9	-3.699

*When p or ρ is less than 0.1%, convert p or ρ to a ratio, set $\eta = \dfrac{p}{1-p}$ (or $\dfrac{\rho}{1-\rho}$), and find the decibel value using a calculator.

Appendices

APPENDIX 1. (Cont.)

P (%)	db	P (%)	db	P (%)	db	P (%)	db	P (%)	db	P (%)	db
30.0	−3.679	35.0	−2.687	40.0	−1.760	45.0	−0.871	50.0	0.000	55.0	0.872
30.1	−3.658	35.1	−2.668	40.1	−1.742	45.1	−0.853	50.1	0.017	55.1	0.889
30.2	−3.637	35.2	−2.649	40.2	−1.724	45.2	−0.835	50.2	0.035	55.2	0.907
30.3	−3.617	35.3	−2.630	40.3	−1.706	45.3	−0.818	50.3	0.052	55.3	0.924
30.4	−3.596	35.4	−2.611	40.4	−1.688	45.4	−0.800	50.4	0.069	55.4	0.942
30.5	−3.576	35.5	−2.592	40.5	−1.670	45.5	−0.783	50.5	0.087	55.5	0.959
30.6	−3.555	35.6	−2.573	40.6	−1.652	45.6	−0.765	50.6	0.104	55.6	0.977
30.7	−3.535	35.7	−2.554	40.7	−1.634	45.7	−0.748	50.7	0.122	55.7	0.995
30.8	−3.515	35.8	−2.536	40.8	−1.616	45.8	−0.730	50.8	0.139	55.8	1.012
30.9	−3.494	35.9	−2.517	40.9	−1.598	45.9	−0.713	50.9	0.156	55.9	1.030
31.0	−3.474	36.0	−2.498	41.0	−1.580	46.0	−0.695	51.0	0.174	56.0	1.047
31.1	−3.454	36.1	−2.479	41.1	−1.562	46.1	−0.678	51.1	0.191	56.1	1.065
31.2	−3.433	36.2	−2.460	41.2	−1.544	46.2	−0.660	51.2	0.209	56.2	1.083
31.3	−3.413	36.3	−2.441	41.3	−1.526	46.3	−0.643	51.3	0.226	56.3	1.100
31.4	−3.393	36.4	−2.423	41.4	−1.508	46.4	−0.625	51.4	0.243	56.4	1.118
31.5	−3.373	36.5	−2.404	41.5	−1.490	46.5	−0.608	51.5	0.261	56.5	1.136
31.6	−3.353	36.6	−2.385	41.6	−1.472	46.6	−0.591	51.6	0.278	56.6	1.153
31.7	−3.333	36.7	−2.366	41.7	−1.454	46.7	−0.573	51.7	0.295	56.7	1.171
31.8	−3.313	36.8	−2.348	41.8	−1.436	46.8	−0.556	51.8	0.313	56.8	1.189
31.9	−3.298	36.9	−2.329	41.9	−1.419	46.9	−0.538	51.9	0.330	56.9	1.206
32.0	−3.273	37.0	−2.310	42.0	−1.401	47.0	−0.521	52.0	0.348	57.0	1.224
32.1	−3.253	37.1	−2.292	42.1	−1.383	47.1	−0.503	52.1	0.365	57.1	1.242
32.2	−3.233	37.2	−2.273	42.2	−1.365	47.2	−0.486	52.2	0.382	57.2	1.260
32.3	−3.213	37.3	−2.255	42.3	−1.347	47.3	−0.468	52.3	0.400	57.3	1.277
32.4	−3.198	37.4	−2.236	42.4	−1.330	47.4	−0.451	52.4	0.417	57.4	1.295
32.5	−3.173	37.5	−2.217	42.5	−1.312	47.5	−0.434	52.5	0.435	57.5	1.313
32.6	−3.153	37.6	−2.199	42.6	−1.294	47.6	−0.416	52.6	0.452	57.6	1.331
32.7	−3.134	37.7	−2.180	42.7	−1.276	47.7	−0.399	52.7	0.469	57.7	1.348
32.8	−3.114	37.8	−2.162	42.8	−1.259	47.8	−0.381	52.8	0.487	57.8	1.366
32.9	−3.094	37.9	−2.144	42.9	−1.241	47.9	−0.364	52.9	0.504	57.9	1.384
33.0	−3.075	38.0	−2.125	43.0	−1.223	48.0	−0.347	53.0	0.522	58.0	1.402
33.1	−3.055	38.1	−2.107	43.1	−1.205	48.1	−0.329	53.1	0.539	58.1	1.420
33.2	−3.035	38.2	−2.088	43.2	−1.188	48.2	−0.312	53.2	0.557	58.2	1.437
33.3	−3.016	38.3	−2.070	43.3	−1.170	48.3	−0.294	53.3	0.574	58.3	1.455
33.4	−2.996	38.4	−2.051	43.4	−1.152	48.4	−0.277	53.4	0.592	58.4	1.473
33.5	−2.977	38.5	−2.033	43.5	−1.135	48.5	−0.260	53.5	0.609	58.5	1.491
33.6	−2.957	38.6	−2.015	43.6	−1.117	48.6	−0.242	53.6	0.626	58.6	1.509
33.7	−2.938	38.7	−1.996	43.7	−1.009	48.7	−0.225	53.7	0.644	58.7	1.527
33.8	−2.918	38.8	−1.978	43.8	−1.082	48.8	−0.208	53.8	0.661	58.8	1.545
33.9	−2.899	38.9	−1.960	43.9	−1.064	48.9	−0.190	53.9	0.679	58.9	1.563
34.0	−2.880	39.0	−1.942	44.0	−1.046	49.0	−0.173	54.0	0.696	59.0	1.581
34.1	−2.860	39.1	−1.923	44.1	−1.029	49.1	−0.155	54.1	0.714	59.1	1.599
34.2	−2.841	39.2	−1.905	44.2	−1.011	49.2	−0.138	54.2	0.731	59.2	1.617
34.3	−2.822	39.3	−1.887	44.3	−0.994	49.3	−0.121	54.3	0.749	59.3	1.635
34.4	−2.802	39.4	−1.869	44.4	−0.976	49.4	−0.103	54.4	0.679	59.4	1.653
34.5	−2.788	39.5	−1.851	44.5	−0.958	49.5	−0.086	54.5	0.784	59.5	1.671
34.6	−2.764	39.6	−1.832	44.6	−0.941	49.6	−0.068	54.6	0.801	59.6	1.689
34.7	−2.745	39.7	−1.814	44.7	−0.923	49.7	−0.051	54.7	0.819	59.7	1.707
34.8	−2.726	39.8	−1.796	44.8	−0.906	49.8	−0.034	54.8	0.836	59.8	1.725
34.9	−2.707	39.9	−1.778	44.9	−0.888	49.9	−0.016	54.9	0.854	59.9	1.743

APPENDIX 1. (Cont.)

P (%)	db	P (%)	db	P (%)	db	P (%)	db	P (%)	db	P (%)	db	P (%)	db
60.0	1.761	65.0	2.688	70.0	3.680	75.0	4.771	80.0	6.021	85.0	7.533		
60.1	1.779	65.1	2.708	70.1	3.700	75.1	4.794	80.1	6.048	85.1	7.567		
60.2	1.797	65.2	2.727	70.2	3.721	75.2	4.818	80.2	6.075	85.2	7.602		
60.3	1.815	65.3	2.746	70.3	3.742	75.3	4.841	80.3	6.102	85.3	7.636		
60.4	1.833	65.4	2.765	70.4	3.763	75.4	4.864	80.4	6.130	85.4	7.671		
60.5	1.852	65.5	2.784	70.5	3.784	75.5	4.888	80.5	6.158	85.5	7.706		
60.6	1.870	65.6	2.803	70.6	3.805	75.6	4.911	80.6	6.185	85.6	7.741		
60.7	1.888	65.7	2.823	70.7	3.826	75.7	4.935	80.7	6.213	85.7	7.776		
60.8	1.906	65.8	2.842	70.8	3.847	75.8	4.959	80.8	6.241	85.8	7.812		
60.9	1.924	65.9	2.861	70.9	3.868	75.9	4.982	80.9	6.269	85.9	7.848		
61.0	1.943	66.0	2.881	71.0	3.889	76.0	5.006	81.0	6.297	86.0	7.884		
61.1	1.961	66.1	2.900	71.1	3.910	76.1	5.030	81.1	6.326	86.1	7.920		
61.2	1.979	66.2	2.919	71.2	3.931	76.2	5.054	81.2	6.354	86.2	7.956		
61.3	1.997	66.3	2.939	71.3	3.952	76.3	5.078	81.3	6.382	86.3	7.993		
61.4	2.016	66.4	2.958	71.4	3.973	76.4	5.102	81.4	6.411	86.4	8.030		
61.5	2.034	66.5	2.978	71.5	3.995	76.5	5.126	81.5	6.440	86.5	8.067		
61.6	2.052	66.6	2.997	71.6	4.016	76.6	5.150	81.6	6.469	86.6	8.104		
61.7	2.071	66.7	3.017	71.7	4.037	76.7	5.174	81.7	6.498	86.7	8.142		
61.8	2.089	66.8	3.036	71.8	4.059	76.8	5.199	81.8	6.527	86.8	8.179		
61.9	2.108	66.9	3.056	71.9	4.080	76.9	5.223	81.9	6.556	86.9	8.217		
62.0	2.126	67.0	3.076	72.0	4.102	77.0	5.248	82.0	6.585	87.0	8.256		
62.1	2.145	67.1	3.095	72.1	4.123	77.1	5.272	82.1	6.615	87.1	8.294		
62.2	2.163	67.2	3.115	72.2	4.145	77.2	5.297	82.2	6.645	87.2	8.333		
62.3	2.181	67.3	3.135	72.3	4.167	77.3	5.322	82.3	6.674	87.3	8.372		
62.4	2.200	67.4	3.154	72.4	4.188	77.4	5.346	82.4	6.704	87.4	8.411		
62.5	2.218	67.5	3.174	72.5	4.210	77.5	5.371	82.5	6.734	87.5	8.451		
62.6	2.237	67.6	3.194	72.6	4.232	77.6	5.396	82.6	6.764	87.6	8.491		
62.7	2.256	67.7	3.214	72.7	4.254	77.7	5.421	82.7	6.795	87.7	8.531		
62.8	2.274	67.8	3.234	72.8	4.276	77.8	5.446	82.8	6.825	87.8	8.571		
62.9	2.293	67.9	3.254	72.9	4.298	77.9	5.471	82.9	6.856	87.9	8.612		
63.0	2.311	68.0	3.274	73.0	4.320	78.0	5.497	83.0	6.886	88.0	8.653		
63.1	2.330	68.1	3.294	73.1	4.342	78.1	5.522	83.1	6.917	88.1	8.694		
63.2	2.349	68.2	3.314	73.2	4.364	78.2	5.548	83.2	6.948	88.2	8.736		
63.3	2.367	68.3	3.334	73.3	4.386	78.3	5.573	83.3	6.979	88.3	8.778		
63.4	2.386	68.4	3.354	73.4	4.408	78.4	5.599	83.4	7.011	88.4	8.820		
63.5	2.405	68.5	3.374	73.5	4.430	78.5	5.624	83.5	7.042	88.5	8.862		
63.6	2.424	68.6	3.394	73.6	4.453	78.6	5.650	83.6	7.074	88.6	8.905		
63.7	2.442	68.7	3.414	73.7	4.475	78.7	5.676	83.7	7.105	88.7	8.948		
63.8	2.461	68.8	3.434	73.8	4.498	78.8	5.702	83.8	7.137	88.8	8.992		
63.9	2.480	68.9	3.455	73.9	4.520	78.9	5.728	83.9	7.169	88.9	9.036		
64.0	2.499	69.0	3.475	74.0	4.543	79.0	5.754	84.0	7.202	89.0	9.080		
64.1	2.518	69.1	3.495	74.1	4.565	79.1	5.780	84.1	7.234	89.1	9.125		
64.2	2.537	69.2	3.516	74.2	4.588	79.2	5.807	84.2	7.267	89.2	9.169		
64.3	2.555	69.3	3.536	74.3	4.611	79.3	5.833	84.3	7.299	89.3	9.215		
64.4	2.574	69.4	3.556	74.4	4.633	79.4	5.860	84.4	7.332	89.4	9.260		
64.5	2.593	69.5	3.577	74.5	4.656	79.5	5.886	84.5	7.365	89.5	9.306		
64.6	2.612	69.6	3.597	74.6	4.679	79.6	5.913	84.6	7.398	89.6	9.353		
64.7	2.631	69.7	3.618	74.7	4.702	79.7	5.940	84.7	7.432	89.7	9.400		
64.8	2.650	69.8	3.638	74.8	4.725	79.8	5.967	84.8	7.466	89.8	9.447		
64.9	2.669	69.9	3.659	74.9	4.748	79.9	5.994	84.9	7.499	89.9	9.494		

APPENDIX 1. (Cont.)

P (%)	db	P (%)	db	P (%)	db	P (%)	db	P (%)	db	P (%)	db	P (%)	db
90.0	9.542	92.0	10.607	94.0	11.950	96.0	13.802	98.0	16.902	100.0	∞ *		
90.1	9.591	92.1	10.666	94.1	12.027	96.1	13.917	98.1	17.129				
90.2	9.640	92.2	10.726	94.2	12.106	96.2	14.034	98.2	17.368				
90.3	9.689	92.3	10.787	94.3	12.186	96.3	14.154	98.3	17.621				
90.4	9.739	92.4	10.840	94.4	12.268	96.4	14.278	98.4	17.889				
90.5	9.789	92.5	10.911	94.5	12.351	96.5	14.405	98.5	18.173				
90.6	9.840	92.6	10.974	94.6	12.435	96.6	14.535	98.6	18.447				
90.7	9.891	92.7	11.038	94.7	12.521	96.7	14.669	98.7	18.804				
90.8	9.943	92.8	11.102	94.8	12.608	96.8	14.807	98.8	19.156				
90.9	9.995	92.9	11.168	94.9	12.697	96.9	14.950	98.9	19.538				
91.0	10.048	93.0	11.234	95.0	12.783	97.0	15.097	99.0	19.956				
91.1	10.111	93.1	11.301	95.1	12.880	97.1	15.248	99.1	20.418				
91.2	10.155	93.2	11.369	95.2	12.974	97.2	15.405	99.2	20.934				
91.3	10.210	93.3	11.438	95.3	13.070	97.3	15.567	99.3	21.519				
91.4	10.264	93.4	11.508	95.4	13.168	97.4	15.736	99.4	22.192				
91.5	10.320	93.5	11.579	95.5	13.268	97.5	15.911	99.5	22.989				
91.6	10.376	93.6	11.651	95.6	13.370	97.6	16.092	99.6	23.962				
91.7	10.433	93.7	11.724	95.7	13.474	97.7	16.282	99.7	25.216				
91.8	10.490	93.8	11.798	95.8	13.581	97.8	16.479	99.8	26.981				
91.9	10.548	93.9	11.873	95.9	13.690	97.9	16.686	99.9	29.996				

APPENDIX 2. ORTHOGONAL ARRAYS AND LINEAR GRAPHS

Introduction

1. The following pages give a collection of standard linear graphs to use in laying out orthogonal arrays for a variety of experiments.
2. "No." indicates the experiment number, and "Col." indicates the column in the orthogonal array.
3. The tables of interactions can be used to find the interactions between two factors in two columns.
4. The groupings of columns in the orthogonal arrays are indicated by the following symbols in the linear graphs. See Chapter 9 in reference (1).

$L_{32}(2^{31})$		$L_{64}(2^{31})$		Other arrays	
Symbol	Group	Symbol	Group	Symbol	Group
○	Groups 1 & 2	○	Groups 1, 2, & 3	○	Group 1
◎	Group 3	◎	Group 4	◎	Group 2
⊙	Group 4	⊙	Group 5	⊙	Group 3
●	Group 5	●	Group 6	●	Group 4

2^n Arrays

$L_4(2^3)$

No. \ Col.	1	2	3
1	1	1	1
2	1	2	2
3	2	1	2
4	2	2	1
	Group 1	Group 2	

Linear graph for L_4

$L_8(2^7)$ Interactions Between Two Columns

Col. No.	1	2	3	4	5	6	7
1	1	1	1	1	1	1	1
2	1	1	1	2	2	2	2
3	1	2	2	1	1	2	2
4	1	2	2	2	2	1	1
5	2	1	2	1	2	1	2
6	2	1	2	2	1	2	1
7	2	2	1	1	2	2	1
8	2	2	1	2	1	1	2

Group 1   Group 2   Group 3

Col. No.	1	2	3	4	5	6	7
(1)		3	2	5	4	7	6
(2)			1	6	7	4	5
(3)				7	6	5	4
(4)					1	2	3
(5)						3	2
(6)							1
(7)							

Linear graph for L_8

(1)   (2)

$L_{12}(2^{11})$

Col. No.	1	2	3	4	5	6	7	8	9	10	11
1	1	1	1	1	1	1	1	1	1	1	1
2	1	1	1	1	1	2	2	2	2	2	2
3	1	1	2	2	2	1	1	1	2	2	2
4	1	2	1	2	2	1	2	2	1	1	2
5	1	2	2	1	2	2	1	2	1	2	1
6	1	2	2	2	1	2	2	1	2	1	1
7	2	1	2	2	1	1	2	2	1	2	1
8	2	1	2	1	2	2	2	1	1	1	2
9	2	1	1	2	2	2	1	2	2	1	1
10	2	2	2	1	1	1	1	2	2	1	2
11	2	2	1	2	1	2	1	1	1	2	2
12	2	2	1	1	2	1	2	1	2	2	1

Group 1   Group 2

Note: The interaction components for two given columns are confounded with the remaining nine columns. Sequential analysis is necessary to find the interactions. This array should therefore not be used for experiments requiring iteractions.

$L_{16}(2^{15})$

No. \ Col.	1	2	3	4	5	6	7	8	9	10	11	12	13	14	15
1	1	1	1	1	1	1	1	1	1	1	1	1	1	1	1
2	1	1	1	1	1	1	1	2	2	2	2	2	2	2	2
3	1	1	1	2	2	2	2	1	1	1	1	2	2	2	2
4	1	1	1	2	2	2	2	2	2	2	2	1	1	1	1
5	1	2	2	1	1	2	2	1	1	2	2	1	1	2	2
6	1	2	2	1	1	2	2	2	2	1	1	2	2	1	1
7	1	2	2	2	2	1	1	1	1	2	2	2	2	1	1
8	1	2	2	2	2	1	1	2	2	1	1	1	1	2	2
9	2	1	2	1	2	1	2	1	2	1	2	1	2	1	2
10	2	1	2	1	2	1	2	2	1	2	1	2	1	2	1
11	2	1	2	2	1	2	1	1	2	1	2	2	1	2	1
12	2	1	2	2	1	2	1	2	1	2	1	1	2	1	2
13	2	2	1	1	2	2	1	1	2	2	1	1	2	2	1
14	2	2	1	1	2	2	1	2	1	1	2	2	1	1	2
15	2	2	1	2	1	1	2	1	2	2	1	2	1	1	2
16	2	2	1	2	1	1	2	2	1	1	2	1	2	2	1
	Group 1	Group 2		Group 3				Group 4							

Interactions Between Two Columns

No. \ Col.	1	2	3	4	5	6	7	8	9	10	11	12	13	14	15
	(1)	3	2	5	4	7	6	9	8	11	10	13	12	15	14
		(2)	1	6	7	4	5	10	11	8	9	14	15	12	13
			(3)	7	6	5	4	11	10	9	8	15	14	13	12
				(4)	1	2	3	12	13	14	15	8	9	10	11
					(5)	3	2	13	12	15	14	9	8	11	10
						(6)	1	14	15	12	13	10	11	8	9
							(7)	15	14	13	12	11	10	9	8
								(8)	1	2	3	4	5	6	7
									(9)	3	2	5	4	7	6
										(10)	1	6	7	4	5
											(11)	7	6	5	4
												(12)	1	2	3
													(13)	3	2
														(14)	1

(1)

a

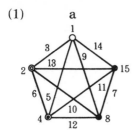

b

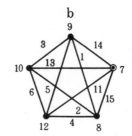

c

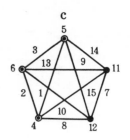

(2)

a

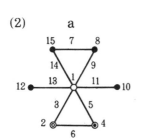

b

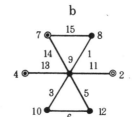

c

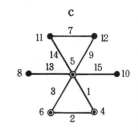

(3)

a

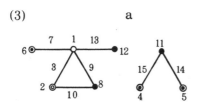

b

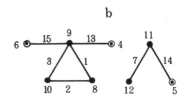

c

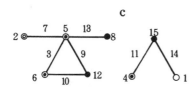

(4)

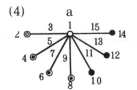 a

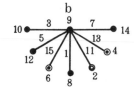 b

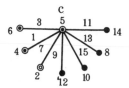 c

(5)

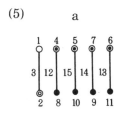 a

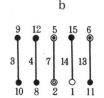 b

 c

(6)

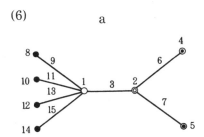 a

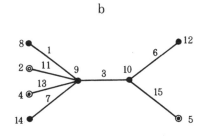 b

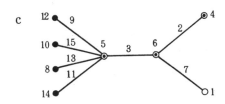 c

3^n Arrays

$L_9(3^4)$

No. \ Col.	1	2	3	4
1	1	1	1	1
2	1	2	2	2
3	1	3	3	3
4	2	1	2	3
5	2	2	3	1
6	2	3	1	2
7	3	1	3	2
8	3	2	1	3
9	3	3	2	1
	Group 1	Group 2		

(1)

$L_{18}(2^1 \times 3^7)$

No. \ Col.	1	2	3	4	5	6	7	8
1	1	1	1	1	1	1	1	1
2	1	1	2	2	2	2	2	2
3	1	1	3	3	3	3	3	3
4	1	2	1	1	2	2	3	3
5	1	2	2	2	3	3	1	1
6	1	2	3	3	1	1	2	2
7	1	3	1	2	1	3	2	3
8	1	3	2	3	2	1	3	1
9	1	3	3	1	3	2	1	2
10	2	1	1	3	3	2	2	1
11	2	1	2	1	1	3	3	2
12	2	1	3	2	2	1	1	3
13	2	2	1	2	3	1	3	2
14	2	2	2	3	1	2	1	3
15	2	2	3	1	2	3	2	1
16	2	3	1	3	2	3	1	2
17	2	3	2	1	3	1	2	3
18	2	3	3	2	1	2	3	1
	Group 1	Group 2			Group 3			

(1)

Interactions can be found without sacrificing columns, by using the two-way layout of columns 1 and 2.

The interactions between three-level columns, however, are partially confounded with the remaining three-level columns.

$$L_{27}\,(3^{13})$$

Col. No. \ No.	1	2	3	4	5	6	7	8	9	10	11	12	13
1	1	1	1	1	1	1	1	1	1	1	1	1	1
2	1	1	1	1	2	2	2	2	2	2	2	2	2
3	1	1	1	1	3	3	3	3	3	3	3	3	3
4	1	2	2	2	1	1	1	2	2	2	3	3	3
5	1	2	2	2	2	2	2	3	3	3	1	1	1
6	1	2	2	2	3	3	3	1	1	1	2	2	2
7	1	3	3	3	1	1	1	3	3	3	2	2	2
8	1	3	3	3	2	2	2	1	1	1	3	3	3
9	1	3	3	3	3	3	3	2	2	2	1	1	1
10	2	1	2	3	1	2	3	1	2	3	1	2	3
11	2	1	2	3	2	3	1	2	3	1	2	3	1
12	2	1	2	3	3	1	2	3	1	2	3	1	2
13	2	2	3	1	1	2	3	2	3	1	3	1	2
14	2	2	3	1	2	3	1	3	1	2	1	2	3
15	2	2	3	1	3	1	2	1	2	3	2	3	1
16	2	3	1	2	1	2	3	3	1	2	2	3	1
17	2	3	1	2	2	3	1	1	2	3	3	1	2
18	2	3	1	2	3	1	2	2	3	1	1	2	3
19	3	1	3	2	1	3	2	1	3	2	1	3	2
20	3	1	3	2	2	1	3	2	1	3	2	1	3
21	3	1	3	2	3	2	1	3	2	1	3	2	1
22	3	2	1	3	1	3	2	2	1	3	3	2	1
23	3	2	1	3	2	1	3	3	2	1	1	3	2
24	3	2	1	3	3	2	1	1	3	2	2	1	3
25	3	3	2	1	1	3	2	3	2	1	2	1	3
26	3	3	2	1	2	1	3	1	3	2	3	2	1
27	3	3	2	1	3	2	1	2	1	3	1	3	2
	Group 1	Group 2						Group 3					

Interactions Between Two Columns

Col. \ Col.	1	2	3	4	5	6	7	8	9	10	11	12	13
	(1)	3	2	2	6	5	5	9	8	8	12	11	11
		4	4	3	7	7	6	10	10	9	13	13	12
		(2)	1	1	8	9	10	5	6	7	5	6	7
			4	3	11	12	13	11	12	13	8	9	10
			(3)	1	9	10	8	7	5	6	6	7	5
				2	13	11	12	12	13	11	10	8	9
				(4)	10	8	9	6	7	5	7	5	6
					12	13	11	13	11	12	9	10	8
					(5)	1	1	2	3	4	2	4	3
						7	6	11	13	12	8	10	9
						(6)	1	4	2	3	3	2	4
							5	13	12	11	10	9	8
							(7)	3	4	2	4	3	2
								12	11	13	9	8	10
								(8)	1	1	2	3	4
									10	9	5	7	6
									(9)	1	4	2	3
										8	7	6	5
										(10)	3	4	2
											6	7	7
											(11)	1	1
												13	12
												(12)	1
													11

(1)

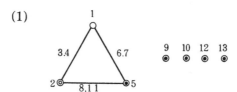

(2)

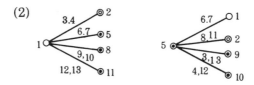

INDEX